AF539387

Fisheries and Aquaculture Technology

Fisheries and Aquaculture Technology

Ahtesham Malik

RANDOM PUBLICATIONS

NEW DELHI - 110 002 (INDIA)

Fisheries and Aquaculture Technology

ISBN 978-93-51112-82-2

Published in 2014 in India by

Reprint 2020

RANDOM PUBLICATIONS

4376-A/4B, Gali Murari Lal, Ansari Road
New Delhi-110 002
Phone: +9111-43580356, 23289044
E-mail: randomexports@gmail.com; sales@randompublications.com; info@randompublications.com

Type Setting by: Friends Media, Delhi-110089
Printed at : Mehra Printers, Delhi-110092

Preface

The systems and technology used in aquaculture has developed rapidly in the last fifty years. They vary from very simple facilities (e.g. family ponds for domestic consumption in tropical countries) to high technology systems (e.g. intensive closed systems for export production). Much of the technology used in aquaculture is relatively simple, often based on small modifications that improve the growth and survival rates of the target species, e.g. improving food, seeds, oxygen levels and protection from predators. Simple systems of small freshwater ponds, used for raising herbivorous and filter feeding fish, account for about half of global aquaculture production. A greater understanding of complex interactions between nutrients, bacteria and cultured organisms, together with advances in hydrodynamics applied to pond and tank design, have enabled the development of closed systems. These have the advantage of isolating the aquaculture systems from natural aquatic systems, thus minimizing the risk of disease or genetic impacts on the external systems. Developments in engineering, some adapted from offshore oil rig construction, increase the possibilities for a progressive offshore expansion of aquaculture using robust cages. Culture-based capture fisheries involving the release of young fish into the wild to improve harvest (an operation also referred to as restocking, stock enhancement or ranching) have existed for a long time for freshwater and anadromous species (e.g. salmon). Sea ranching, however, has just made a start but its long-term viability is being assessed. Advances have also been made in capture-based aquaculture involving the growing/fattening of young fish (e.g. tuna) captured from the wild. Potential conflicts with capture fisheries are being assessed. Major progress have also being made in the aquafeeds technology, combining a large number of ingredients into very small pellets.

For both capture fisheries and aquaculture, the technological development and widespread use of synthetic fibres, hydraulic equipment for gear and fish handling, electronics for fish finding, satellite-based technology for navigation and communications, onboard conservation and increased use of outboard engines have all contributed to the major expansion of fisheries and aquaculture in recent decades - particularly in small-scale fisheries. Technical advances have generally led to more efficient and economical fishing operations, reduction of the physical labour required per unit of output and improved access to resources. Where management has been ineffective, the greater efficiency of fishing methods and aquaculture production has sometimes led to overfishing and environmental degradation. This points to the need to develop more effective fisheries management frameworks, together with safer and more environmentally-friendly methods of production, for example, in developing selective fishing gear and in designing aquaculture systems that reduce their impact on external environments.

This book is an attempt in this direction and hope that will be useful to students, academicians and researchers.

I thank all members of my team who have helped in the preparation of the book. My special thanks go to "Random Publications" who have published the book.

—Ahtesham Malik

Contents

1

Aquaculture Methods and Practices

Historical Perspective

Aquaculture has a tradition of about 4 000 years. It began in China, possibly due to the desires of an emperor to have a constant supply of fish. It is speculated that the techniques for keeping fish in ponds originated in China with fishermen who kept their surplus catch alive temporarily in baskets submerged in rivers or small bodies of water created by damming one side of a river bed. Another possibility is that aquaculture developed from ancient practices for trapping fish, with the operations steadily improving from trapping-holding to trapping-holding-growing, and finally into complete husbandry practices. Chinese who emigrated to other Southeast Asian countries probably carried the knowledge with them and inspired the local people to take up fish farming. Brackishwater aquaculture is thought to have originated in Indonesia with the culture of milkfish and grey mullet (Ling, 1977) and must have spread to neighbouring countries like the Philippines which has been practising it for about 300 to 400 years. The husbandry of fish is therefore not a new phenomenon. Ancient practices based on the modifications of natural bodies of water or wetlands to entrap young fish in enclosures until harvest, have just evolved into more systematic and scientific methods and techniques.

Other regions of the world have shorter traditions of aquaculture. In North America, it is about a century old; in Africa, aquaculture production consists almost exclusively of tilapia culture in freshwater ponds and dates back to the 1940s (UNDP/NORAD/FAO, 1987). Aquaculture development has been very recent and is just gaining momentum in Australia, New Zealand, and the Pacific Island countries (Rabanal, 1988b).

Table: *Possible Environmental Impacts of Aquaculture*

Culture System	*Environmental Impact*
EXTENSIVE	
1. Seaweed culture	May occupy formerly pristine reefs; rough weather losses; market competition; conflicts/failures, social disruption.
2. Coastal bivalve culture (mussels, oysters, clams, cockles)	Public health risks and consumer resistance (microbial diseases, red tides, industrial pollution; rough weather losses; seed shortages; market competition especially for export produce; failures, social disruption.
3. Coastal fishponds (mullets, milkfish, shrimps, tilapias)	Destruction of ecosystems, especially mangroves; increasingly non-competitive with more intensive systems; nonsustainable with high population growth; conflicts/failures, social disruption.
4. Pen and cage culture in eutrophic waters and/or rich benthos (carps, catfish, milkfish tilapias)	Exclusion of traditional fishermen; navigational hazards; conflicts, social disruption; management difficulties; wood consumption.
SEMI-INTENSIVE	
1. Fresh- and brackishwater pond (shrimps and prawns, carps, catfish, milkfish, mullets, tilapias)	Freshwater: health risks to farm workers from waterborne diseases. Brackishwater: salinization/acidification of soils/aquifers. Both: market competition, especially for export produce; feed and fertilizer availability/prices; conflicts/failures, social disruption.
2. Integrated agriculture-aquaculture (rice-fish; live stock/poultry-fish; vegetables - fish and all combinations of these)	As freshwater above, plus possible consumer resistance to excreta-fed produce; competition from other users of inputs such as livestock excreta and cereal brans; toxic substances in livestock feeds (e.g., heavy metals) may accumulate in pond sediments and fish; pesticides may accumulate in fish.
3. Sewage-fish culture (waste treatment ponds; latrine wastes and septage used as pond inputs; fish cages in wastewater channels)	Possible health risks to farm workers, fish processors and consumers; consumer resistance to produce.
4. Cage and pen culture, especially in eutrophic waters or on rich benthos (carps, catfish, milkfish, tilapias)	As extensive cage and pen Systems above.
INTENSIVE	
1. Freshwater, brackishwater and marine ponds (shrimps; fish, especially carnivores - catfish, snakeheads, groupers, sea bass, etc.)	Effluents/drainage high in BOD and suspended solids; market competition, especially for export product; conflicts/failures, social disruption.
2. Freshwater, brackishwater and marine cage and pen culture (finfish, especially carnivores - groupers, sea bass, etc. - but also some omnivores such as common carp)	Accumulation of anoxic sediments below cages due to fecal and waste feed build-up; market competition, especially for export produce; conflicts/failures, social disruption; consumption of wood and other materials.
3. Other - raceways, silos, tanks, etc.	Effluents/drainage high in BOD and suspended solids; many location-specific problems.

Source: Modified from Pullin, 1989

A number of aquaculture practices are used world-wide in three types of environment (freshwater, brackishwater, and marine) for a great variety of culture organisms. Freshwater aquaculture is carried out either in fish ponds, fish pens, fish cages or, on a limited scale, in rice paddies. Brackishwater aquaculture is done mainly in fish ponds located in coastal areas. Marine culture employs either fish cages or substrates for molluscs and seaweeds such as stakes, ropes, and rafts. Culture systems range from extensive to intensive depending on the stocking density of the culture organisms, the level of inputs, and the degree of management. In countries where government priority is directed toward increased fish production from aquaculture to help meet domestic demand, either as a result of the lack of access to large waterbodies (e.g., Nepal, Central African Republic) or the over-exploitation of marine or inland fisheries (e.g., Thailand, Zambia), aquaculture practices are almost exclusively oriented toward production for domestic consumption (UNDP/NORAD/FAO, 1987).

These practices include:

(i) freshwater pond culture;

(ii) rice-fish culture or integrated fish farming;

(iii) brackishwater finfish culture;

(iv) mariculture involving extensive culture and producing fish/ shellfish (e.g., oysters, mussels, cockles) which are sold in rural and urban markets at relatively low prices.

Extensive systems use low stocking densities (e.g., 5 000-10 000 shrimp post larvae (PL)/ha/crop) and no supplemental feeding, although fertilization may be done to stimulate the growth and production of natural food in the water. Water change is effected through tidal means, i.e., new water is let in only during high tide and the pond can be drained only at low tide. The ponds used for extensive culture are usually large (more than two ha) and may be shallow and not fully cleared of tree stumps. Production is generally low at less than 1 t/ha/y.

Semi-intensive systems use densities higher than extensive systems (e.g., 50 000-100 000 shrimp PL/ha/crop) and use supplementary feeding. Intensive culture uses very high densities of culture organism (e.g., 200 000-300 000 shrimp PL/ha/crop) and is totally dependent on artificial, formulated feeds. Both systems use small pond compartments of up to one ha in size for ease of management.

Table: *Aquaculture Production Systems and Practices*

Region	***Major Culture Species***	***Major Culture Systems***	***Major Culture Practices***	***Scope for Future Development/Needs for Further Expansion***
ASIA	At least 75 species; diverse freshwater and marine species, including high-value shrimps, molluscs, seaweeds, with carps and seaweeds dominating production	Traditional extensive to intensive	- Fish ponds - Fish pens and fish cages - Floating rafts, lines, and stakes for molluscs and seaweeds	Development of culture-based fisheries in inland lakes, rivers, floodplains, and permanent and temporary reservoirs and barrages
				Resource enhancement programmes integrated with environmental management
PACIFIC	Mussels and oysters, red seaweeds	Intensive/semi-intensive to extensive	- Hanging lines for mussels and pearl oysters	Production of high-value species for select markets;
			- Offshore cages for salmon	Small-scale aquaculture for local markets;
			- Pond culture for shrimps, tilapia, catfish, milkfish	Improved management of fishery resources, particularly reef fisheries
			- Freshwater pens for crayfish	
LATIN AMERICA	50 species of fish, crustaceans, and molluscs, including freshwater fish and marine shrimps in South America and molluscs in Central America	Extensive to semi-intensive and Intensive	- Offshore cage farming of Pacific and Atlantic salmon - Ocean ranching in Southern Ocean - Semi-intensive farming of marine shrimp in coastal	Production of species for export and marine shrimp and salmon

Source: ADCP Aquaculture Regional Profiles

Semi-intensive and intensive culture systems are managed by the application of inputs (mainly feeds, fertilizers, lime, and pesticides) and the manipulation of the environment primarily by way of water management through the use of pumps and aerators. Feeding of the stock is done at regular intervals during the day.

In intensive shrimp culture, the computed daily feed ration is given in equal doses from as low as three to as high as six times a day. Water change is also effected on a daily basis, with approximately 10-15% of the water in the pond replenished by the entry of new water in semi-intensive shrimp ponds.

Semi-intensive and intensive culture systems are therefore more labour-intensive than extensive systems which need little attention, and are costlier to set up and operate, not to mention the fact that they also carry higher risks of mortalities resulting from disease, poor management, and/or force majeure (e.g., from anoxia due to non-functioning aerators during times of power failure).

Production is of course much higher (for example, ranging from a minimum of 1.5 t/ha/crop from semi-intensive shrimp culture to a high of 10 t/ha/crop from intensive shrimp culture). Financial returns are therefore much more attractive than those from extensive culture, although studies have shown that the return on investment (ROI) from semi-intensive culture is better than from intensive culture due to the high cost of inputs (largely fry and feeds) used in intensive culture.

Fish Pond Culture

Pond culture, or the breeding and rearing of fish in natural or artificial basins, is the earliest form of aquaculture with its origins dating back to the era of the Yin Dynasty (1400-1137 B.C.). Over the years, the practice has spread to almost all parts of the world and is used for a wide variety of culture organisms in freshwater, brackishwater, and marine environments. It is carried out mostly using stagnant waters but can also be used in running waters especially in highland sites with flowing water.

Running water fish culture involves growing the fingerlings to marketable size in earthen ponds using water from rivers, irrigation canals, or plain rain water. The system approximates intensive culture in that it involves the application of rapid water changes and the heavy stocking of the cultured species. The continuously flowing water is advantageous for fish culture as it supplies abundant dissolved oxygen and flushes away waste products and unconsumed feeds.

Table: *Summary of Comparative Features among the Three Main Culture Systems*

Parameter	***Extensive***	***Semi-Intensive***	***Intensive***
Species Used	Monoculture or Polyculture	Monoculture	Monoculture
Stocking Rate	Moderate	Higher than extensive culture	Maximum
Engineering Design and Layout	May or may not be well laid-out	With provisions for effective water management	Very well engineered system with pumps and aerators to control water quality and quantity
	Very big ponds	Manageable-sized units (up to 2 ha each)	Small ponds, usually 0.5-1 ha each
	Ponds may or may not be fully cleaned	Fully cleaned ponds	Fully cleaned ponds
Fertilizer	Used to enhance natural productivity	Used regularly with lime	Not used
Pesticides	Not used	Used regularly for prohylaxis	Used regularly for prophylaxis
Food and Feeding Regimen	None	Regular feeding of high quality feeds	Full feeding of high-quality feeds
		Depending on stocking density used, formulated feeds may be used partially or totally	
Cropping Frequency (crops/y)	2	2.5	2.5
Quality of Product	Good quality	Good quality	Good quality
	Culture species dominant but extraneous species may occur	Confined to culture species	Confined to culture species
	Variable sizes	Uniform sizes	Uniform sizes

Culture Species

Commonly raised species in freshwater ponds are the carps, tilapia, catfish, snakehead, eel, trout, goldfish, gouramy, trout, pike, tench, salmonids, palaemonids, and the giant freshwater prawn Macrobrachium. In brackishwater ponds, common species include milkfish (Chanos chanos), mullet (Mugil sp.) and the different penaeid shrimps (Penaeus monodon, P. orientalis, P. merguiensis, P. penicillatus, P. semisulcatus, P. japonicus, and M. ensis). The more popular species for culture in marine ponds are the sea bass, grouper, red sea bream, yellowtail, rabbitfish, and marine shrimps.

In Asia, where the bulk of world production from aquaculture emanates, fish ponds are mostly freshwater or brackishwater, and rarely marine. In China and most of the Indian sub-continent, pond culture is traditionally dominated by freshwater species, mainly the carps, usually in polyculture and/or integrated with animal husbandry. In Southeast Asia, fish ponds are predominantly brackishwater, with milkfish and penaeid shrimps grown either in polyculture or in monoculture. Recently in Latin America and the Caribbean, brackishwater pond culture of penaeid shrimps has expanded rapidly, as it has in some parts of Asia.

In Africa, the tilapias and carps dominate aquaculture production. Controlled breeding is also carried out in ponds with goldfish, trout, Bagrus and, to a lesser extent, Lates niloticus, Heterotis niloticus, and Clarias lazera. Ten species of molluscs belonging to four genera (Crassostrea, Mytilus, Venerupis and Pinctada) are cultured. Crustacean culture has yet to be developed on a significant scale (Satia, 1989).

Site Selection

Proper site selection is recognised as the first step guaranteeing the eventual success of any aquaculture project and forms the basis for the design, layout, and management of the project (SCSP, 1982a). For fish ponds, especially those to be used for coastal/brackishwater aquaculture of high-value species like shrimps, site selection is critical and should be given utmost attention.

Adisukresno (1982), Hechanova (1982), and Jamandre and Rabanal (1975) listed the following guidelines for the selection of a suitable site for coastal fish ponds:

(i) *Soil Quality:* preferably, clay-loam, or sandy-clay for water retention and suitability for diking; alkaline pH (7 and above) to prevent problems that result from acid-sulphate soils (e.g., poor fertilizer response; low natural food production and slow growth of culture species; probable fish kills).

(ii) *Land Elevation and Tidal Characteristics:* preferably with average elevation that can be watered by ordinary high tides and drained by ordinary low tides; tidal fluctuation preferably moderate at 2-3 m. (Sites where tidal fluctuation is large, say 4 m, are not suitable because they would require very large, expensive dikes to prevent flooding during high tide. On the other hand, areas with slight tidal fluctuation, say 1 m or less, could not be drained or filled properly.)

(iii) *Vegetation:* preferably without big tree stumps and thick vegetation which entail large expense for clearing; areas near river banks and those at coastal shores exposed to wave action require a buffer zone with substantial growths of mangrove. (The presence of Avicennia indicates productive soil; nipa and trees with high tannin content indicate low pH.)

(iv) *Water Supply and Quality:* with steady supply of both fresh and brackish water in adequate quantities throughout the year; water supply should be pollution-free and with a pH of 7.8-8.5.

(v) *Accessibility:* preferably readily accessible by land/water transport; close to sources of inputs such as fry, feeds, fertilizers, and markets, fish ports, processing plants, and ice plants; and linked by communication facilities to major centres.

(vi) Availability of manpower for construction and operation.

Pond Layout

The layout of the pond system depends on the species for culture and on the size and shape of the area, which in turn determines the number and sizes of ponds and the position of the water canals and gates. A fish farm is considered properly planned if all the water control structures, canals, and the different pond compartments mutually complement each other (SCSP, 1982a). A complete fish farm has nursery and grow-out ponds and, in some instances, transition ponds for intermediate-sized fish/shrimp, all of which are properly proportioned and positioned within.

Milkfish culture in brackishwater ponds in the Philippines follows the traditional practice of providing for nursery, transition, and rearing operations.

In some cases, formation ponds are used for additional growth or stunting of fingerlings prior to stocking in rearing ponds. The nursery ponds comprise about 1-4% of the total production area while the transition and formation ponds constitute about 6-9% of total area (Camacho and Lagua, 1988).

It has been suggested that a similar progressive culture scheme be adopted for shrimp pond culture when no supplementary feeding is practised.

For growing to a medium size, a two-stage progression composed of a nursery pond (NP) and a rearing pond (RP) is adequate; for growing to larger sizes, a three-stage progression composed of nursery, transition, and rearing ponds is recommended (ASEAN/SCSP, 1978).

In general, however, shrimp monoculture uses direct stocking of post larvae in rearing ponds and therefore requires only one type of pond with separate inlets and outlets for better circulation and aeration.

Design of Pond Facilities

A fish pond system consists of the following basic components:

(i) pond compartments enclosed by dikes;

(ii) canals for supply and drainage of water to and from the pond compartments; and

(iii) gates or water control structures to regulate entry and exit of water into and from the pond compartments.

Pond compartments are usually rectangular in shape although in Indonesia, running water ponds are generally triangular, raceway-shaped, or oval. They vary in size from less than a hectare to several hectares each, sometimes up to 20-50 ha in size. However, with the new intensive methods, the trend is to use smaller units for flexibility and ease of management.

The elevation of the rearing pond bottom for milkfish is usually such that only a maximum of 40 cm of water can be held in the ponds during the culture period (Jamandre and Rabanal, 1975). For new shrimp ponds, the minimum water depth is 1 m.

The entire pond system is enclosed by a perimetre dike and the individual pond compartments are separated from each other by partition dikes. The outer perimetre dike is usually wider and higher than the inner partition dikes and serves to protect the entire fish pond area from flooding and destruction brought about by tide and wave action. The inner dikes are narrower and shorter.

The design of the dikes depends primarily on soil characteristics. Dikes are usually earthen although intensive shrimp ponds are concrete-lined or brick-lined as in Taiwan (PC). The side slopes are designed for structural stability, the ratio of horizontal length to height ranging from 1:1 to 1:3. The height and width of dikes depend on the type (primary, secondary, or tertiary), tide conditions, flood level, pond water depth, soil shrinkage, and freeboard (SCSP, 1982a).

The following slopes are recommended for dikes built with good clay soil:

- 2:1 when dike height is above 4.26 m and exposed to wave action;

- 1:1 when dike height is less than 4.26 m and tidal range is greater than 1 m; and
- 1:2 when tidal range is 1 m or less, and dike height is less than 1m.

The dike crown should not be less than 0.5 m and the main dike surrounding the farm should be 0.5 m above the highest dike or flood level recorded in the locality (ASEAN/SCSP, 1978).

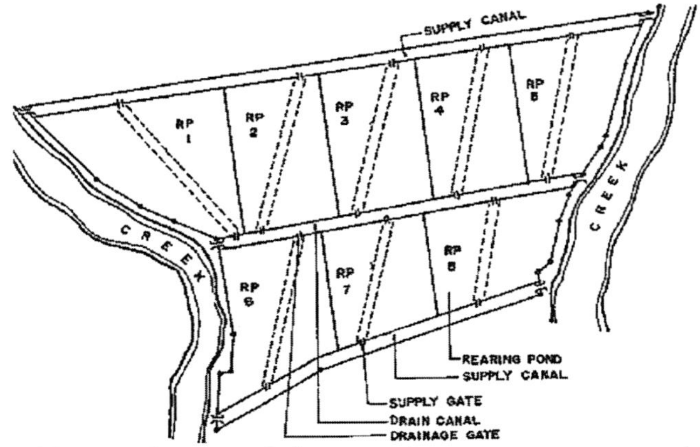

Figure: *Pond layout showing shrimp pond compartments, canals, and gates.*

Fig. 6. Typical cross sections of dikes. (A)

Water conveyance structures (canals/channels) supply new water into the pond and drain out old water. They also provide the facility for holding and harvesting of fish and of serving as waterways for transporting farm supplies. Traditional milkfish ponds usually have only one canal that is used for both supply and drainage. Shrimp ponds have separate supply and drainage canals. Canals which are to be used for harvesting should be 30 cm below the level of the pond bottom to allow draining of pond water.

Having separate water intake and discharge canals in a pond complex brings about the following advantages (ASEAN/SCSP, 1978):

(i) Better filling and non-contamination of pond by discharge from other ponds.

(ii) Greatly reduced possibility of spread of disease.

(iii) Maintenance of constant head in intake canal thus reducing water loss through leaks/seepages in pond dikes and consequently reducing leaching of acids into the ponds from dikes with acid-sulphate soils.

(iv) Absence of conflict of usage between farmers.

(v) Better water exchange for individual ponds, and

(vi) Possibility of effecting flow-through systems.

The width of the canals depends on the amount of water they must carry. The following should be taken into account when designing canals:

(i) Volume of water to be held in the ponds.

(ii) Time requirement for filling or draining the pond.

(iii) Amount of rainfall which must be carried off in a given period of time.

(iv) Elevation of canal bottom in relation to tide.

(v) Other uses like transportation, harvesting of milkfish, and holding of broodstock (ASEAN/SCSP, 1978).

Diversion canals are constructed where there is much runoff from adjoining areas, to prevent sudden salinity changes and the possible entry of polluted, pesticide-loaded water and/or of silted water into the pond complex (Jamandre and Rabanal, 1975).

The entry and exit of water into ponds through the canals is regulated or controlled by gates. Main gates regulate the exchange of water between the pond system and the tidal stream or sea, and may be constructed of reinforced concrete or wood. Reinforced concrete is more expensive but lasts longer. Such a gate has one or multiple (2, 3, 4, etc.) openings depending on the relative size of the pond unit to be served. A recent innovation for a smaller and less expensive main gate is the monk-type gate which uses culverts usually made of concrete hollow blocks. The SEAFDEC Aquaculture Department has also introduced the open sluice gate made of ferro-cement (Corre, 1988).

Secondary gates, which regulate water exchange between the ponds and the canals, are usually made of wood. Pipes or culverts can also be used for smaller ponds such as nursery or fry ponds and transition ponds for milkfish culture. Secondary gates are now usually located toward one end of the narrower side of the pond compartment to give good turbulence and circulation during the filling and draining.

Shrimp ponds are provided with separate supply and drainage gates to effect flow-through water management and facilitate water exchange through supply and drainage canals (NACA, 1986). Inlet and outlet gates are best located at opposite corners of the same pond

(ASEAN/SCSP, 1978), across which a diagonal trench, about 5-10 m wide and 0.3-0.5 m deep, extending from inlet to outlet gates is recommended for convenient draining of water.

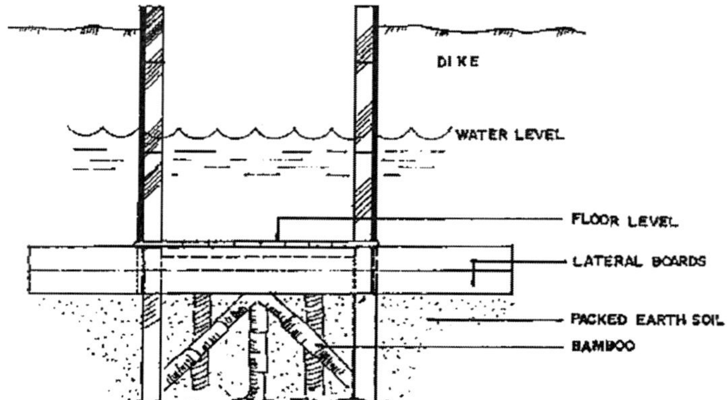

***Figure:** Diagram of wooden gate.*

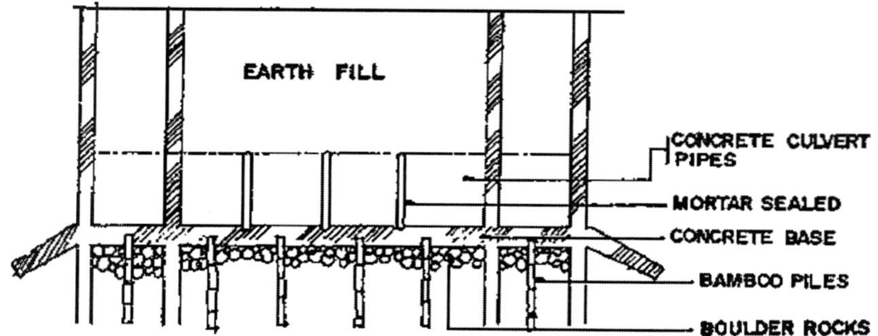

***Figure:** Use of culvert pipes as secondary gates*

Gates should be located where they are not exposed to strong weather forces and where water of good quality can be allowed to enter the fish pond system. Proper gate location can also serve to aerate the pond water and promote water circulation (SCSP, 1982a).

During the construction of gates for shrimp ponds a number of requirements should be kept in mind (ASEAN/SCSP, 1978), and the gates should:

- be durable, water-tight, and made of locally available materials;
- have adequate capacity for the amount of water to be taken in or drained;
- allow water to be taken in or discharged at the bottom;
- have provisions for draining pond surface water;

- have gate bottom elevation that permits complete draining of pond water;
- have slots or grooves for the placement of outside and inside screens to prevent undesirable species from entering the pond and the shrimps from leaving the pond;
- have place for net installation for harvesting; and
- be easy to operate.

Pond Management

Pond management techniques for finfish and shrimp culture, while varying slightly depending on the specific biological requirements of the culture organism, the type of culture system, and the culture environment (freshwater, brackishwater, and marine), are similar in that they involve the following basic activities:

- Pond preparation/conditioning.
- Stocking.
- Feeding and/or fertilization (depending on the culture system used).
- Water management.
- Pond maintenance, and
- Harvesting.

Variations would consist mainly of differences in application rates of fertilizers, lime, pesticides, and feeds; stocking rates and sizes of stocking material; rate of water change; and harvesting techniques.

Extensively managed systems generally require the least management, with no supplemental feeding and minimal water exchange on account of the low stocking density used. On the other hand, intensively managed ponds require full artificial feeding and substantial water management to ensure optimum culture conditions for the species being reared.

Pond Preparation

Ponds are totally drained and the pond bottoms dried prior to the application of pesticides. Tobacco dust, derris root/rotenone powder, teaseed cake/powder, or Gusathion-A are used to eliminate predators and/or wild species that may eventually compete with the cultured organisms for food and space. Teaseed cake is perhaps the best fish poison to use in brackishwater ponds to selectively kill unwanted fish without damaging the shrimps and without affecting rotifers and copepods which are feed for shrimps. On the other hand, rotenone is most effective in fresh water and works better in low-salinity water.

Table: *Variations in pond management techniques commonly*

Species	***Stocking Rate***	***Fertilization***	***Feed Type***	***Rate of Water Change***	***Pesticides/Predator Control***		***Reference***
					Type	***Application Rate***	
MILKFISH (Chanos chanos)	2 000-5 000/ha	16-20-0 at 50 kg/ha; 45-0-0 at 15 kg/ha; chicken manure at 0.5 t/ha twice weekly	Rice bran and trash fish as supplemental feed	Once every two weeks at high tide	Lime; ammonium sulfate	1 t/ha 10 g/m²	Bombeo - Tuburan & Gerochi, 1988
TILAPIA (O. niloticus; O. mossambicus)	5 000-20 000/ha	Chicken manure at 500 kg/ha; Inorganic fertilizers at 50 kg/ha	Rice bran, fish meal, ipil-ipil leaf meal				Camacho & Lagua, 1988
CATFISH (Clarias botrachus and monocephalus)	60-300/m²		9 parts trash fish and 1 part rice by-products	When necessary			Sirikul et al., 1988
PENAEIDS	From as low as 15 000 to as high as 300 000/ha	Chicken manure at 1-2 t/ha followed by inorganic fertilizer at 75-150 kg/ha mono-ammonium phosphate (16-20-0) and 25-50 kg/ha of urea (46-0-0)	Supplemental feed of rice bran with trash fish, mussels, and clam meat; artificial/formulated diets with 40% CP.	20-30% once every week or every two weeks for low density ponds; 5-20% daily for semi-intensive to intensive ponds			Corre, 1988

Ponds with acid-sulphate soils are repeatedly dried and flushed, i.e., filled and drained to remove the acids formed by pyrite oxidation.

Agricultural lime is then applied to correct soil pH and bring it up to at least 6.5. Brackishwater ponds are usually treated by spreading 1.5 t of agricultural lime per ha, followed by another 1.5 t worked into the soil.

To stimulate and maintain the growth of natural plankton, organic (e.g., chicken manure) or inorganic fertilizer (e.g., urea, ammonium phosphate) are applied to the pond bottom. After fertilizer application, water is let in to a depth of about 20-40 cm and gradually increased to 1 m a week after fertilization. Intensively managed ponds or ponds where artificial feeding shall be given, do not need to be fertilized. Extensive ponds need regular fertilization during the culture period to maintain the growth of natural food. Semi-intensive ponds may use a mix of fertilization and supplementary feeding.

Stocking

After the pond is prepared, fish fingerlings or shrimp post larvae are stocked at the appropriate density depending on the culture strategy, size of pond, and the size of fingerlings, among others.

The fingerlings are properly acclimated and conditioned prior to stocking and weak or diseased fish eliminated. Stocking is usually done in the early morning or late afternoon.

Feeding

Fish/shrimp grown in semi-intensive and intensive culture ponds are given supplementary and full artificial feeds, respectively, the former to augment the natural food in the pond, the latter to totally replace the natural organisms in the water as a source of nutrition.

A wide variety of feed ingredients is used to prepare supplemental/ artificial feeds. The simplest fish feeds are prepared at the pond site using locally available raw materials like rice or corn bran, copra meal, and rice mill sweepings as sources of carbohydrates. These are usually mixed with animal protein like trash fish/fish meal, shrimp heads, and snail meat. Supplemental feeds for tilapia are prepared using 80% rice bran and 20% fish meal. Those for shrimps in improved extensive culture (low-density stocking but given dietary supplements for increased growth/production) usually include fresh raw materials like snail/mussel/clam meat or carabao hide and other slaughterhouse leftovers.

Commercial feed preparations are also available now in a wide range of brandnames, mostly for semi-intensive and intensive shrimp culture. (Taiwan (PC), Japan, and the USA are the top producers of commercial fish/shrimp feeds.) These commercial diets consist of a number of ingredients like fish meal, blood meal, bone meat, and shrimp head meal (to serve as attractant for the shrimp), together with vitamin and mineral premix and carbohydrate sources like rice/ corn bran or wheat.

The crude protein (CP) content of these shrimp feeds is generally not lower than 30% to satisfy the high animal protein requirement of shrimps, actually estimated to be about 40% during the earlier stages of growth.

Commercial feeds usually come in various formulations to match the protein requirement of the culture organism, which as a rule, decreases with age. Thus, fish/shrimp feeds come in different forms as starter, grower, and finisher, with starter feeds having the highest CP content of about 40% and finisher feeds having the lowest CP content of about 20%. Starter feeds are usually given on the first month of culture, finisher feeds on the last month, and grower feeds in between.

Some shrimp culturists prefer not to give artificial feeds during the first two weeks of culture when the newly stocked post larvae can subsist on the plankton available in the water.

The feeding rate is computed as a percentage of the estimated animal biomass in the pond, with higher rations given when the animals are small and gradually decreasing as they become bigger. The daily feeding rate usually starts at 5% and 10-15% of estimated biomass of fish and shrimps, respectively, and decreases to a low of 2% and 5%, for fish and shrimps, respectively, toward harvest.

The daily feed rations are given in equal portions during the course of a day. Freshwater fish like tilapia are usually fed twice a day - early morning and late afternoon. Penaeid shrimps are fed more frequently, from three to four to as often as six to seven times a day.

Feeds are broadcast into the water and/or supplied on feeding trays. In semi-intensive and intensive shrimp ponds, small feeding boats are used by caretakers who go around the pond distributing the feed by broadcasting. At certain points along the periphery of the pond, feeding trays are submerged into the water after known quantities of feed are put on the surface, to supply feed to the shrimps in the pond as well as to monitor feed consumption and shrimp growth. The feeding tray is lifted two to three hours after the feed was supplied to check how much of it has been consumed and to see if the shrimps are healthy and feeding. Empty feeding trays may indicate that the quantity given is inadequate and may have to be increased. Conversely, full or slightly touched trays indicate excessive feed quantities and/or sluggish shrimps. The feeding ration is subsequently adjusted accordingly to optimize feed utilisation.

By monitoring the feeding tray, one can get a good indication of the sizes and quantity of shrimps present in the pond without a need for cast-netting or actual sampling, since shrimps are invariably found on the tray when it is lifted out of the water.

Water Management

Water in the pond is kept at certain levels for optimal fish growth. In general, a pond water depth of 1 metre is considered best for culture of tilapia, carps, and shrimps; traditional milkfish ponds can do with just 40-60 cm of water.

Pond water is not just maintained at a certain depth; its quality must also be kept high to ensure optimal growth of the culture organism. This is particularly important in semi-intensive and intensive culture systems where large amounts of metabolites are continously excreted into the pond and where excess, unconsumed feeds add to the bottom load and serve to pollute the water.

To prevent the deterioration of the pond environment, pond water is continuously freshened by the entry of new water from the river or water source (through the supply canal) while old water is drained through the outlet/drainage gate and through the drainage canal into the sea or river.

A flow-through system of water management that allows the simultaneous entry and exit of water into and out of the pond is essential in any high-density culture system. This is effected by the provision of separate inlets and outlets for all the ponds, each inlet regulating the flow of water from the supply canal to the pond and each outlet controlling the discharge of water out of the pond into the drainage canal. Both the supply and drain gates are so designed as to bring water into and drain water out of the lower levels of the pond, where water quality tends to get poorer faster as a result of the accumulation of wastes and their subsequent decomposition.

The regular replenishment of pond water, independent of natural tidal fluctuations, is made possible by the use of pumps which draw water from the source even at low tide. Although there is no hard-and-fast rule as to the rate of water change necessary for medium- to high density aquaculture, semi-intensive culture systems usually change water at the rate of 10% daily for an equivalent total replacement of water every ten days or three times per month.

Intensively managed ponds require greater water exchange in view of the much higher organic load on the pond bottom, especially toward the latter part of the culture cycle when the animals excrete more wastes.

Intensive ponds/tanks usually need to provide for aeration facilities/ equipment to prevent anoxia that may lead to mass mortalities. Oxygen depletion in high-density ponds results not only from the faster rate of utilisation of dissolved oxygen for respiratory activities; it is also caused by the fast rate of decomposition at the pond bottom by aerobic or oxygen-consuming micro-organisms.

Paddlewheels or other types of aerators are thus provided in the ponds to effect the infusion/introduction of greater quantities of oxygen into the water and prevent fish/shrimp mortalities. The aerators are usually operated at regular/periodic intervals for certain fixed durations during the day but especially in the early morning hours when the concentration of dissolved oxygen is known to be lowest (as a result of the absence of photosynthetic, oxygen-producing activity in the pond). Toward the end of the culture period when oxygen demand is highest, aeration may have to be provided continuously and not just sporadically as could be done during the initial stages of rearing. At that time too, water pumps usually need to be run for longer periods to effect greater water exchange.

Pond water is also regularly sampled and measurements taken of basic/essential parameters, particularly dissolved oxygen, pH, and salinity. This is important for the purpose of determining the need for corrective/remedial action to bring water quality to optimum levels and obtain good yields.

Dissolved oxygen levels are kept, as much as possible, above 5 ppm by pumping and aeration. Problems of acidity are corrected by liming. Salinity is an important parameter for penaeid culture and has to be maintained within a range of 15-25 ppt for best results.

During summer months, high-salinity water can be diluted by mixing with fresh water from springs or deep wells.

Pond Maintenance

(i) *Fertilization:* Aside from feeds and water management, the following pond maintenance procedures are carried out: regular application of fertilizers, lime, and pesticides; prevention of entry of predators; monitoring of the stock for growth rate determination as a basis of feeds and water management; and regular pond upkeep and maintenance. Extensive ponds are fertilized regularly using either organic fertilizers like chicken, cow, or pig manure, or inorganic fertilizers like urea, ammonium phosphate, or both, to maintain the plankton population in the pond. The fertilizers are either broadcast over the pond water surface or kept in sacks suspended from poles staked at certain portions along the pond periphery. Semi-intensive and intensive culture systems do not require fertilization since they are not natural food-based, except for those which grow plankton-feeders like milkfish whose diet is largely algae dependent.

(ii) *Liming:* In addition to fertilization, ponds also need to be given regular doses of lime to maintain water pH at alkaline or near-alkaline levels (preferably not lower than six). Agricultural lime is broadcast over the pond and applied on the sides of the dikes to correct soil and water acidity.

(iii) *Elimination of Pests and Predators:* Unwanted and predatory species which may have survived the application of pesticides during pond preparation or which were able to enter the pond through the gate screens or through cracks in the dikes, are eliminated by the application of pesticides, preferably organic, into the pond. Crabs, which are a serious problem in shrimp ponds because they are carnivorous and cause damage to the pond dikes, are not usually affected by known pesticides and are therefore best eliminated by the use of crab traps situated in the pond. It is also important that the gates are properly screened and the screens kept whole, to prevent the entry of small unwanted fish into the pond. Double screens are usually installed at the main intake to ensure that pests and predators are prevented from entering the pond system.

(iv) *Stock Monitoring:* The culture organisms are monitored closely and regularly to determine their rate of growth and the general condition of the stock. They are regularly sampled for length-weight measurements as a basis for determining/estimating their biomass in the pond and therefore their daily feed rations, as well as for making projections on harvest schedules and procurement of pond inputs. In the first few months of culture, the feeding tray is a good tool for stock monitoring. As the organisms grow in size, cast-netting is used as a sampling tool, with those caught in the throw of the cast net providing an indication as to sizes and weights of stock. Based on the sampled weights and the daily feed consumption, it is possible to predict the available biomass (i.e., stock surviving after initial mortalities) and make projections on volume of harvest. For this purpose, it is essential that accurate records are kept for analysis at a later time. Data on initial size/weight and number of fry/post larvae stocked, average body weight at each sampling, and feed consumption on a daily basis, are important to have on file.

(v) *Regular Upkeep and Maintenance of Facilities:* The pond dike and gates are checked regularly for cracks that could lead to seepages and losses of stock. The dikes are best planted with grass or vegetative cover to prevent erosion. The gates and other support infrastructure are properly maintained for efficient operation.

Harvesting

Marketable-size fish/shrimps are harvested at the end of the culture period by draining the pond and using harvesting nets to catch the fish or shrimps. The latter are harvested with a bag-net attached to the sluice gate as water is drained out of the pond at low tide. Tilapia are harvested using seine nets after the pond water is drained to half-level the night before.

Harvest of milkfish takes advantage of their behaviour of swimming against the current. The method, known in the Philippines as "pasulang" or "pasubang" involves draining 85-90% of the pond water during low tide and allowing in the water at the incoming high tide so that the fish swim against the current through the tertiary gate and into the catching pond, whose gate is closed once a large number of fish is impounded. The fish in the catching pond are then harvested by seining and the rest hand-picked.

Integrated Fish Farming

In a number of countries in Asia (e.g., China, Nepal, Thailand, Malaysia, Indonesia) and in some parts of Africa, freshwater fish culture is integrated with the farming of crops, mainly rice, vegetables and animals (usually pigs, ducks, and chickens). This leads to greater overall efficiency of the farming system as wastes/by-products or one component are used as inputs in another. For example, poultry or pig manure can be used to fertilize the fish pond and the vegetable garden and the waste vegetables can be fed to the fish and the pigs.

In Africa, fish culture in rice fields and in combination with pig and duck rearing, is not too widely practised but has significant potential. Reported fish yields ranged from 2 000-4 000 kg/ha/y with ducks, 8 500-8 900 kg/ha/y with pigs, and 3 600-4 900 kg/ha/y with poultry in Gabon. It has also been proven economically viable since it involves minimal investment. Its spread has, however, been constrained by the widespread use of pesticides in many countries (Satia, 1989).

Pen and Cage Culture

Pen and cage culture involve the rearing of fish within fixed or floating net enclosures supported by frameworks made of bamboo, wood, or metal, and set in sheltered, shallow portions of lakes, bays, rivers, and estuaries.

Compared to fish pond culture with its 4000-year tradition, fish pen/cage culture is of more recent origin. Cage culture seems to have developed independently in at least two countries - in Kampuchea where fishermen in and around the Great Lake region would keep Clarias spp. and other commercial fishes in bamboo or rattan cages and baskets; and in Indonesia where bamboo cages have been used to grow Leptobarbus hoeveni fry as early as 1922. Since then, cage culture has spread throughout the world to more than 35 countries in Europe, Asia, Africa, and the Americas (Beveridge, 1984).

Pen culture is said to have originated in the Inland Sea area of Japan in the early 1920s (Alferez, 1977), adopted by the People's Republic of China in the 1950s for rearing carps in freshwater lakes (Beveridge, 1984), and introduced to culture milkfish in the shallow, freshwater, eutrophic Laguna de Bay in the Philippines in the 1970s (Baguilat, 1979). From there it has been successfully extended for the culture of tilapia and carps (Rabanal, 1988b). Its development and adoption as a popular technology has not been widespread, though, perhaps because of its site-specific requirements like its suitability mainly in shallow lentic environments. At present, it is commercially practised only in the Philippines, Indonesia, and China.

The wider popularity of cage culture as compared to pen culture may be due to its greater flexibility in terms of siting the structures. For example, cages may be installed in bays, lagoons, straits, and open coasts as long as they are protected from strong monsoonal winds and rough seas. Floating cages can also be set up in deep lakes and reservoirs, and in rivers and canal systems, and even in deep mining pools which could not be used otherwise for culture due to harvesting difficulties.

In general, however, both pen and cage culture have expanded rapidly, especially over the past two decades vis-a-vis the decreasing availability of land-based resources for fish culture and an increasing awareness of their merits over traditional pond culture, such as:

(i) their applicability in different types of open water bodies like coastal waters, protected coves and bays, lakes, rivers, and reservoirs;

(ii) their high productivity (of as much as 10-20 times that of ponds Of comparative sizes) with minimal inputs and at lower costs to develop and operate; and

(iii) the greater socio-economic opportunities they provide to low-income families in the rural areas, particularly those displaced by the reduction of fish catches in over-exploited coastal, municipal waters, because they require comparatively low capital outlay and use simple technology.

Yields from pen and cage culture are generally high, with or without supplemental feeding depending on the natural productivity of the water body. In the Philippines, for example, the yields of milkfish from fish pens in Laguna de Bay were as high as 4 t/ha/crop (compared to a national milkfish fish pond average of 1 t/ha/y in 1980 when the productivity of the lake was very high at 1 700 mg C/m^3/hr (Baluyut, 1983).

In Indonesia, the cage culture of common carp in the Lido Reservoir in Cigombong gave a total production of 28 kg/m^2 at a stocking density of 6 kg/m^2 (Baluyut, 1983). The cage culture of marine finfishes has likewise been shown to give high yields.

Culture Species

The choice of species for stocking and rearing in pens and cages is governed by much the same criteria as in species selection for pond culture, including (Guerrero, 1982):

(i) fast growth in confinement;

(ii) good consumer acceptance;

(iii) high tolerance to a wide range of environmental conditions;

(iv) resistance to disease;

(v) ready supply of fish seed for stocking; and (vi) ease of culture and management.

Table: *Comparison of Production of Cage-Cultured Marine Fish*

Species	***Seriola T: quinqueradiata***	***Trachinotus carolinus***	***Polydactylus sexfilis***	***Epinephelus salmoides****
Country of culture	Japan	Florida, USA	Hawaii, USA	Penang, Malaysia
Initial stocking density				
fish/m^3	10	250	50	60
kg/m^3	0.15-0.55	1.75	0.4	3.4
Rearing period (days)	225	273	300	240
Production (kg/m^3)	0.85-14.45	44.7	-	41.4
Average production rate (kg/m^3/day)	0.004-0.06	0.16	-	0.17
Mean size of fish				
Initial (g)	10-50	7	9	55.7
At harvest (g)	1 000-2 000	213.6	300	795.9
Average growth rate (g/fish/day)	4.40-8.67	0.76	0.97	3.08

Source: *SEAFDEC/IDRC, 1979*

There are approximately ten species of fish which are commercially cultured in cages and pens in both temperate and tropical waters, including tilapias (S. mossambicus and S. niloticus); carps (Chinese, Indian, and common varieties); milkfish; snakeheads and catfishes; marble goby; and salmonids (rainbow trout, salmon). Marine species include mainly grouper, sea bass, mullet, snapper, and milkfish.

In the Philippines, Indonesia, and China, pen culture is limited to the following species: milkfish (Chanos chanos); tilapia; and the Chinese carps: bighead, silver carp, grass carp (Ctenophanyngodon idella); and common carp. Other species have been suggested as possible candidates for utilisation in pen/cage culture in the following three different environments:

(i) *Freshwater:* Habitats with high natural productivity (e.g., lakes, oxbow lakes, swamps, mining pools, rivers, and reservoirs): mullets, eels, catfish, Puntius gonionotus. Habitats with low natural productivity: Leptobarbus, Clarias batrachus, Oxyeleotris, and Macrobrachium.

(ii) *Brackishwater:* Sea bass, mullet, siganids, sea bream, grouper, snapper, threadfin, carangids. Hilsa spp., Sparus spp., and eels.

(iii) *Marine:* Siganids, pampano, yellowtail, tuna, grouper, snapper, sea bass, sea bream, carangids, pomfret.

Site Selection

The selection of sites for fish pen/cage culture should be guided by the following basic criteria:

Table: *Commercially Important Species in Inland Water Cage and Pen Farming*

Species		*Countries*	*Climate*	*Type of feeding*	*Lotic/Lentic*	*Cage/Pen*
Salmonids	Rainbow trout	Europe, North America, Japan, high altitude tropics (eg Colombia, Bolivia, Papua New Guinea)	Temperate	Intensive. High protein (40%)	Lentic	Floating cage
	Salmon (various species) smolts	Europe, North America, South America, Japan	Temperate	Intensive. High protein (452)	Lentic	Floating cage
Carps	Chinese carps (Silver carp, grass carp, bighead carp)	Asia, Europe, North America	Temperate - tropical	Mainly semi-intensive, although also extensive (Asia) and intensive (Europe North America)	Lotic and lentic	Cages and pens
	Indian major carps (Labeo rohita)	Asia	Sub-tropical -tropical	Semi-intensive	Mainly lentic	Mainly cages
	Common carp	Asia, Europe, North America, South America	Temperature -tropical	Mainly semi-intensive, although also intensive	Mainly lentic	Mainly cages
Tilapias	(O. Mossambicus, O. niloticus, etc.)	Asia, Africa, North America, South America	Sub-tropical -tropical	Mainly semi-intensive, although also intensive	Mainly lentic	Mainly cages
Catfishes	Channel catfish	North America	Temperature -sub-tropical	Intensive	Lentic	Floating cages
	Clarias spp.	Southeast Asia, Africa	Tropical	Semi-intensive	Lotic and lentic	Floating cages
Snakeheads	Channa spp. Ophicephalus spp.	Southeast Asia	Tropical	Semi-intensive/intensive	Lotic and lentic	Floating cages
Pangasius spp.		Southeast Asia	Tropical	Semi-intensive	Lentic	Floating cages
Milkfish		Southeast Asia	Tropical	Semi-intensive	Lentic	Pens

Source: Beveridge, 1984

(i) Protection from high winds or typhoons.

(ii) Adequate water exchange that will enable the flow of nutrient-laden water through the pens/cages.

(iii) Good water quality (high or adequate dissolved oxygen, stable pH, and low turbidity, and absence of pollution).

(iv) Firm bottom mud to allow pen framework to be driven deep into substrate for better support.

(v) Freedom from predators and natural hazards.

(vi) Accessibility to sources of inputs, including labour and markets, and

(vii) Good peace and order condition.

The factors to be considered in selecting sites for pens and cages in freshwater, brackishwater, and marine environments. It is important to note that the selection of a suitable site is vital to the success of the culture system; a good site selected solves much of the management problems of pen/cage culture (Chua, 1979).

Design and Construction

Both fish pens and fish cages are built around the same basic design concept: a net enclosure supported by a rigid framework. They differ, however, in a number of respects. Firstly, a pen does not have a net bottom; the edges of its net wallings/fencings are anchored to the lake bottom/substrate by means of bamboo pegs and the lake bottom is the pen bottom. In comparison, a cage is like an inverted mosquito net with the cage bottom made of the same netting material used for its four sides.

Secondly, fish pens theoretically have no limit to their size/area while cages cannot exceed 1 000 m^2 in area for reasons of the quantity of material required for cage construction (due to the need for a flooring) and manageability of operation (cages have to be lifted and the fish scooped out and not harvested using nets as in pens).

Thirdly, design of the structures and methods of construction are different. Fish pens are fixed structures; fish cages may either be fixed or floating. Fish pens for milkfish culture in Laguna de Bay, Philippines consist of a nursery pen within the grow-out pen/enclosure. Cages are individual units for either seed production or grow-out; they are, however, usually installed in clusters or modules with a common framework.

Pens and cages come in various shapes and sizes and are made of different types of materials. Most pens and cages are rectangular or square although some may be circular, as in some milkfish pens in Laguna de Bay and the milkfish broodstock cages at the SEAFDEC Aquaculture Department in the Philippines, or cylindrical as those used for fish collection in Malaysian or Indonesian fresh waters. Rectangular cages are preferred for easy operation and management. Circular cages are more suitable for some species like milkfish and yellowtail but are more expensive to build (SEAFDEC/IDRC, 1979).

Table: *Factors to be Considered in the Selection of Cage/Pen*

	Marine	*Brackishwater*	*Freshwater*
Protection from Elements			
Natural	Wind direction	Water current	Wind direction
	Lagoons, bays and coves offer	Erosion and	Water current
	differing	accretion	Floods
	situations	siltation	Typhoons
Artificial	Breakwaters	Deflectors	Breakwaters
Water Circulation			
Related to protection	Currents	Currents	Currents
	Tidal levels	Tidal levels	Stratification and up-welling
Net pen spacing	Well-spaced	Well-spaced	Well-spaced
Water Quality and Soil Type			
Chemical	Salinity	Salinity	Soil type
	Type of bottom	Type of bottom	pH, NH3, BOD, hardness
		Pesticides and fertilizer run-off	Saltwater intrusion
Physical	Temperature	Temperature	Temperature
	Siltation and turbidity	Siltation and turbidity	Siltation and turbidity
	Tidal fluctuation	Tidal fluctuation	Depth fluctuation
	Topography	Topography	Texture of the substratum
		Floating objects	Topography
			Floating objects
Biological	Predators, pests and competitors	Predators, pests and competitors	Algal bloom
	Vegetation	Vegetation	Predators, pests and competitors
	Plankton bloom	Plankton and benthos	Vegetation
	Diseases and parasites	Diseases and parasites	Diseases and parasites
			Natural productivity
Pollution	Industrial pollutants	Industrial pollutants	Industrial pollutants
			Thermal pollution
	Domestic pollutants	Domestic pollutants	Agricultural pollutants
	Agricultural pollutants	Agricultural pollutants	Mine pollution
Access and Security			
Supplies	Materials	Materials	Materials
	Feeds	Feeds	Feeds
	Fingerlings	Fingerlings	Fingerlings
Markets (live and fresh sales)	Close to market	Close to market	Close to market
Labor	Availability	Availability	Availability
	Cost	Cost	Cost
Monitoring	Easy access necessary for regular monitoring visits.		
Security	Efficient precautions and security from interference of all sorts.		
Others	Frequency of navigation	Frequency of navigation	Frequency of navigation
	Property rights, policies and laws	Property rights, policies and laws	Property rights, policies and laws
	Social aspects	Social aspects	Social aspects

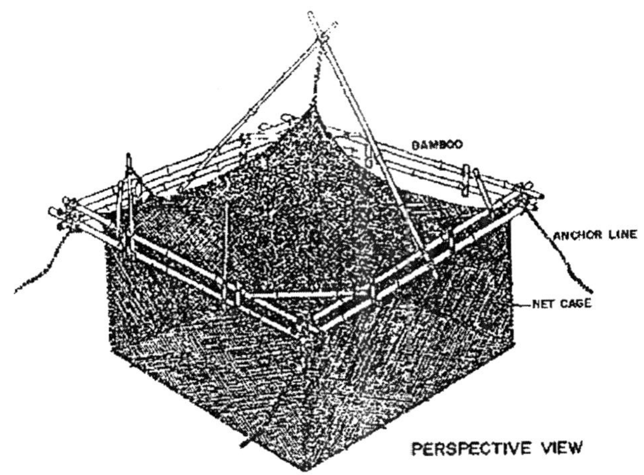

Figure: *Perspective view and parts of a floating cage. (A)*

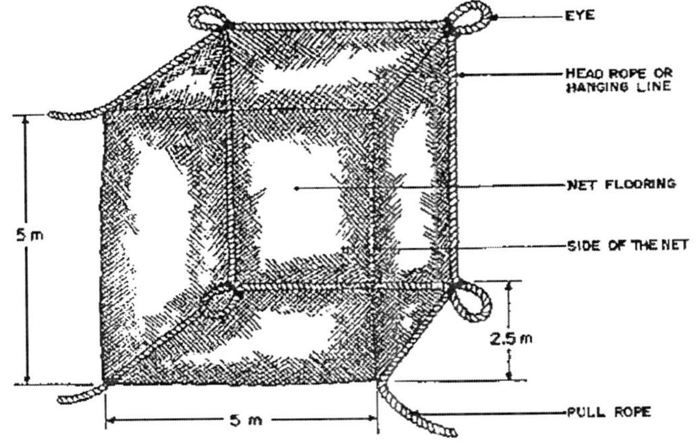

Figure: *Perspective view and parts of a floating cage. (B)*

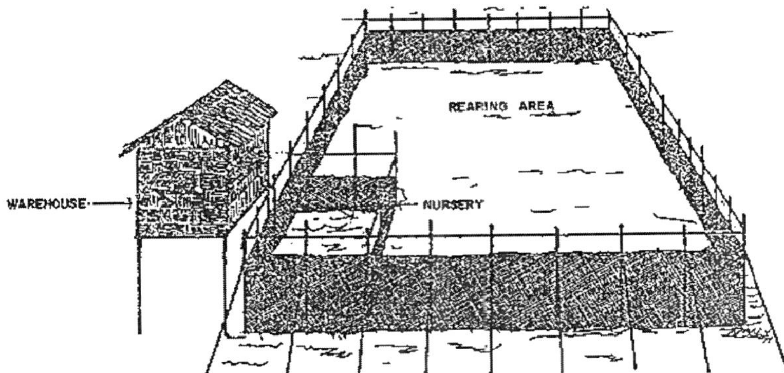

Figure: *Perspective of a fishpen showing nursery pen within the grow-out enclosure.*

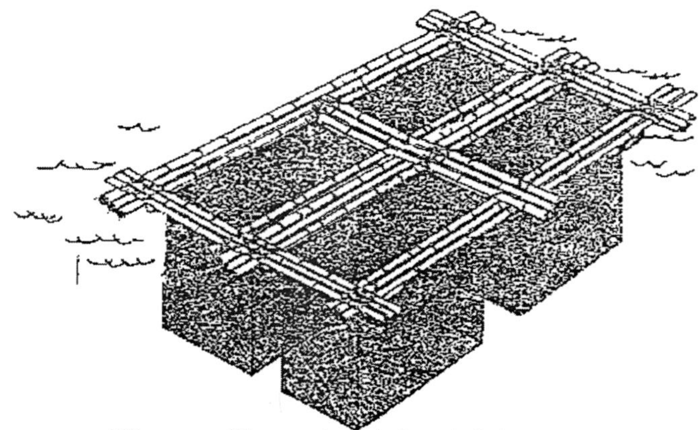

Figure: *Cluster/module of fish cages.*

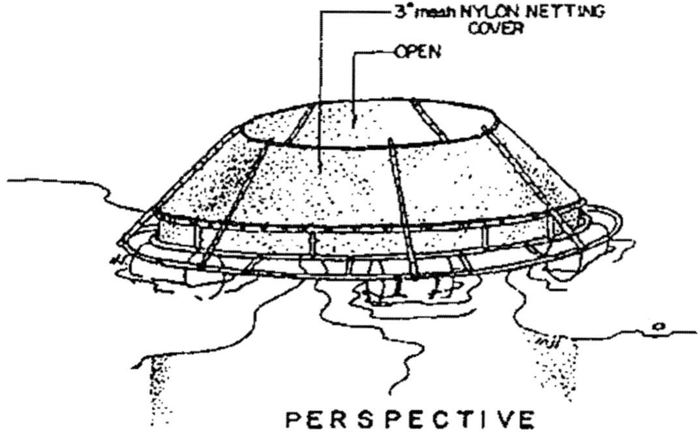

Figure: *Circular milkfish broodstock cage used at the SEAFDEC Aquaculture Department.*

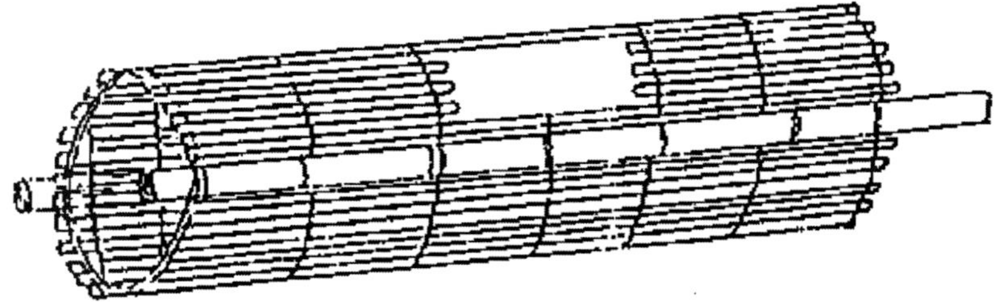

Figure: *Cylindrical fish cage made of bamboo and rattan*

Polyethylene and nylon monofilament twine are widely used for fabricating cages and net pens although wire mesh is used in several countries. The framework structure is generally made out of bamboo and other locally available wood. Cage floatation materials include bamboo, PVC pipes/containers, steel or plastic drums, styrofoam, and aluminum floats. The type of anchor for floating cages varies depending on the depth of water, nature of bottom, tides, and currents. Concrete slabs of different sizes and shapes, sand bags, and iron anchors are widely used in different countries.

Pen and Cage Operation

Basic procedures involved in the management of pen and cage culture are very much like those in pond culture, starting with completion of construction and preparation of the culture facilities for stocking, rearing, and harvesting.

Slight variations in specific activities exist, however, as the result of the very nature of the system. For example, it is obviously not possible to apply fertilizers, lime, and pesticides since the system has open water exchange between the inner compartment and the outside environment.

Soon after construction of the pen/cage is completed, preparations are made to procure fry/fingerlings for stocking. Milkfish pens have a nursery compartment into which milkfish fry are grown for 3-4 weeks to 12 cm long fingerlings which can be released into the grow-out compartment.

The nursery pen and the grow-out compartment are prepared for stocking by clearing the bottom of predatory fish like Megalops cyprinoides and Elops hawaiiensis. The milkfish fry/fingerlings from the nursery pen are stocked in the rearing pen at 20 000-50 000 per ha where they are cultured to marketable size.

In the Philippines, the milkfish stock in the pen is not generally given supplemental feeding except for occasional rations of bread crumbs, rice bran, broken ice cream cones, fish meal, and ipil-ipil leaf mill.

On the other hand, cage-reared fish may or may not be fed supplemental or artificial diets depending on the stocking density used and the level of technology in the country. Cage feeding trials in the Philippines showed the adequacy of a ration composed of 77% rice bran and 23% fish meal with feed conversion ratios of 2.2-2.8. Current feed practices in freshwater cage culture involve the provision of supplemental feeds using readily available ingredients like rice bran and poultry feeds. Other countries use artificial feeds based on simple diets preferably prepared in pelleted form for best results.

At the end of the culture period, the fish are harvested from pens using harvesting nets (e.g., gill nets, cast nets, seines) or from cages by lifting the cage and causing the fish to collect in one corner for scooping out using a pail.

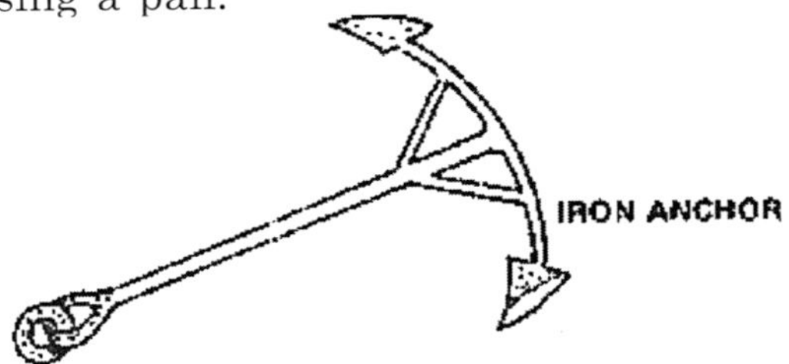

Figure: *Types of anchor used for floating cages (from SEAFDEC/IDRC, 1979). (A)*

Figure: *Types of anchor used for floating cages (from SEAFDEC/IDRC, 1979). (B)*

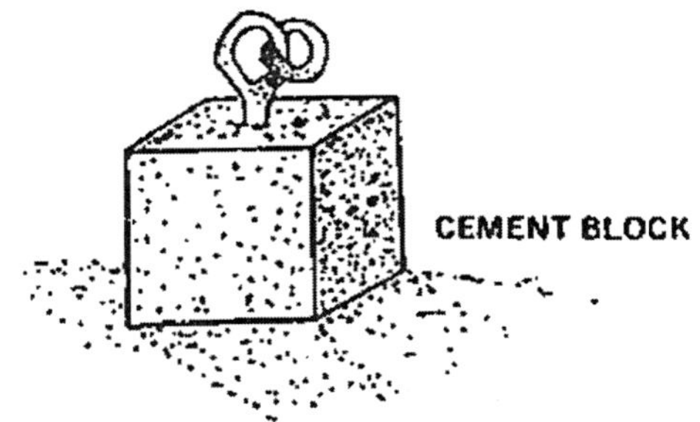

Figure: *Types of anchor used for floating cages (from SEAFDEC/IDRC, 1979). (C)*

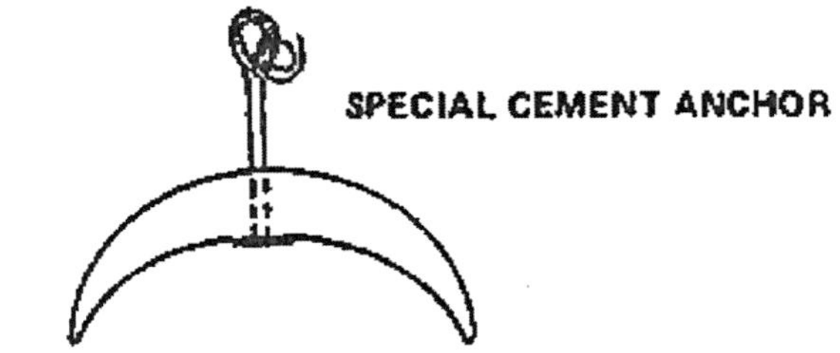

Figure: *Types of anchor used for floating cages (from SEAFDEC/IDRC, 1979). (D)*

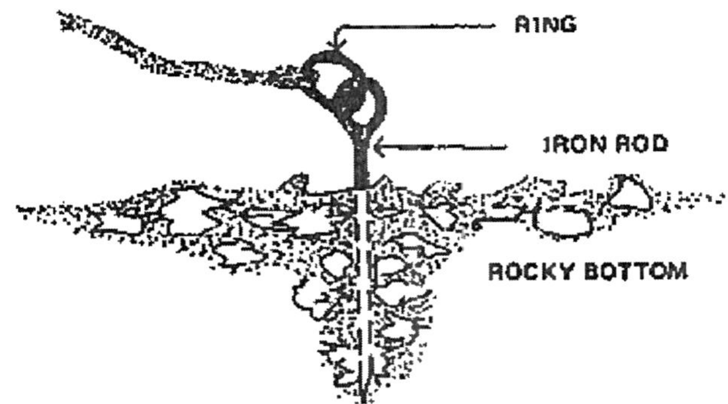

Figure: *Types of anchor used for floating cages (from SEAFDEC/IDRC, 1979). (E)*

Table: *Feed types given to cage-reared fish*

Country	*Culture Species*	*Feed Type*	*Reference*
GDR	Common carp	Formulated feed/pellets, 33.7% CP	Muller, 1979
USSR	Common carp	Mixture of minced trash fish, molluscs, crayfish, and grown cereals	- do -
Hungary	Wels (Silurens glanis)	Trash fish, slaughterhouse wastes, cereal grain meals	- do -
-do-	Carp polyculture (common, silver, bighead)	Pelleted common carp feed	
India	Indian carp polyculture	Soya bean powder, ground nut, oil cake, rice polish (1:1.1)	Natarajan et al., 1979
Indonesia	Leptobarbus hoeveni and Thynnichthys thynoides	Coconut water, cassava, rubber leaves	Reksalegora, 1979
Indonesia	S. niloticus	Aquatic plants (Lemna, Hydrila, Chara)	Rifai, 1979
Nepal	Common carp	Wheat flour, rice bran, mustard oil cake	Sharma, 1979
Thailand	Catfish, sand goby, common carp, local carp, tilapia, snakehead	Pellets consisting of ground fish meal, soy bean, peanut, and rice bran	Tangtrongpiros, 1979
	Sea bass (Lates calcarifer)	Trash fish	Dhebtaranon et al., 1979

Open Water Culture

The farming of molluscs and seaweeds in open marine waters has become increasingly popular in a number of countries, especially in the Third World where it is seen as a viable alternative to municipal or artisanal fisheries or as a means of supplementary income for small-scale fishermen. Because seafarming is generally low-cost and labour-intensive and could thus involve entire coastal communities, it is particularly appropriate in areas where production from municipal fisheries has substantially declined and where, as a result, subsistence fishermen have little or no means of livelihood.

Mollusc Culture

Bivalves are widely cultured in a number of countries world-wide. In Asia and the Pacific, they represent a high quality food resource with annual production higher than from crustacean culture on a per hectare basis (Sitoy, 1988). In 1984, molluscs accounted for approximately 35% of the total production of coastal aquaculture in terms of gross weight in the region (Shang, 1986).

The most important species for culture in Southeast Asia are the oysters (mainly Crassostrea spp.), mussels (mainly Perna spp.), clams, cockles, and scallops (Pagcatipunan, 1987; Sitoy, 1988; Cheong, 1988; Liong et al., 1988).

In Japan, the most commonly cultured species include Crassostrea gigas, C. rivularis, C. nippona, C. echinata, and Ostrea denseramellosa, with C. gigas as the predominant species (Honma, 1980). In Africa, the culture of Venerupis is reported in Tunisia and Pinctada spp. in Sudan (Shehedah, 1975). In Mexico, the culture of the large oyster Crassostrea spp. is carried out by cooperative societies and of the mussel Mytilus edulis on floating rafts by private investors.

Oysters are widely distributed in estuaries and bays which receive some run-off from land and have somewhat lower salinity than the open sea. As they filter their food from the water, they grow best in areas with moderate to high concentrations of phytoplankton (SCSP, 1982c). Oysters grow best in intertidal areas where they are exposed for some minutes or a few hours during low tide (Pagcatipunan, 1987). Mussels, on the other hand, cannot tolerate tidal exposure even during low tide.

The best sites for culturing molluscs are therefore those that meet their biological requirements, including the following:

(i) Seawater salinity range of 15-35 ppt.

(ii) Water depth of 1-10 m, and

(iii) Muddy bottom for mussels and hard rocky or coralline substrates for oysters.

In addition, the area for mollusc culture should be protected from strong water currents reaching three knots and should be accessible to source of seed, transport, and markets. Furthermore, the presence of local available stock in an area is a good indicator of its suitability for mollusc culture.

Countries which have successfully cultured bivalve molluscs have developed their own systems of culture which depend entirely on natural seed stock, which are either gathered from natural seed beds or collected using suitable materials for collecting seed from natural grounds.

In the Philippines, both natural and synthetic ropes have been used for spat collection. However, since natural ropes, which have been found to attract more larvae than synthetic polyethylene or polypropylene ropes, do not last long, natural fibrous materials like coconut coir are sometimes interwoven with synthetic nylon ropes to make them more attractive to the larvae.

The string seed collectors are submerged in the sea water for seed collection at the right time. They are hung on a collector rack, normally 12 strings along a distance of 1.8 m to hold about 1 000 shells. Sometimes, strings are hung separately from each other at regular intervals; at others, three or four strings are put together for hanging to prevent branches from attaching to strings when they occur in large quantities.

Three principal methods of oyster culture are used in the Philippines and Japan: (i) hanging methcd including rafts, longlines, simple hanging, and rocks; (ii) stake or stick method; and (iii) broadcast or sowing method (SCSP, 1982c; Honma, 1980).

In Japan, the earliest method used at the Hiroshima Prefecture, where oyster culture began in the 17th century, was the stick culture method. In 1927, the hanging method of culture was introduced which later developed into different variations, viz., the simple hanging method, raft method, and longline method, to suit different local conditions as culture grounds shifted from inner to outer parts of the bay to outer open seas.

The broadcast system is actually used throughout the world in places where the bottom of shallow bays is firm enough to support the materials used as collectors and for growing oysters. Oyster shells, stones, or other hard objects are scattered on the bottom in areas where setting or the attachment of oyster larvae is known to occur. The young oysters or spat are left in places attached to the collectors until they are large enough for harvest (SCSP, 1982c).

The stake method is usually applied in shallow areas with soft or muddy bottom, usually not more than 1 m deep during low tide. The stakes, usually bamboo trunks (whole or split), branches of mangrove trees, or concrete Y-shaped posts and other similar materials are staked on the sea bottom in rows spaced about 0.5 m apart, to serve as attachment for oyster spat.

The hanging method of oyster culture uses empty oyster shells or other material such as coconut shells as collectors. The collectors are strung on synthetic twine or heavy monofilament nylon, and placed about 10 cm apart by using bamboo tubes as spacers or by tying knots in the twine. The strings are hung from a platform or rack/tray made of bamboo or wooden splits or welded wire with wooden frame, and placed on wooden plots. Oysters detached from the collectors or those small oysters/seedlings which are separated from harvested stocks are cultured on the trays until they are big enough for the market. Harvesting procedures vary with the culture method. Oysters grown on stakes or by hanging are removed from the stakes or ropes on shore or in a boat after the stakes/ropes are lifted out of the water. Those grown by broadcasting are usually collected at low tide.

Mussel farming makes extensive use of bamboos either as stakes or as floating rafts. The stake method, similar to that for oyster culture, is the most commonly used. The mussels are harvested by divers after 6-10 months when they reach a length of 5-8 cm.

Alternatively, mussels are grown on floating rafts which have the following advantages: (i) faster growth; (ii) possibility of regular thinning and therefore higher production per unit area; (iii) possibility of transfer to other areas to prevent siltation; and (iv) ease of construction using more durable materials (Sitoy, 1988).

Mussels and oysters grown in waters contaminated by domestic and industrial wastes need to undergo depuration or cleansing, using artificially cleaned water or clean seawater from saltwater wells, to ensure satisfactory microbiological and chemical quality of the product.

Seaweed Farming

Seaweeds, aside from being used as food, are important sources of colloids or gels, such as agar, as well as minerals of medicinal importance such as iodine. Eucheuma, a red algae, is a valuable source of carrageenan, an important industrial compound used in stabilising and improving the quality of a great number of products. Caulerpa lentillifera, a green algae, is economically important because it is a favourite and nutritious salad dish containing essential trace minerals such as calcium, potassium, magnesium, sodium, copper, iron and zinc. It is also known for its medicinal properties, being used as an anti-fungal agent and as a natural means for lowering blood pressure. Gracilaria, another red alga, is economically important in Taiwan (PC) for its agar extracts.

Table: *Species Under Cultivation in the Asia-Pacific Region*

Seaweed Groups/Species	*Country Where Cultivated*
A. Green Seaweeds (Chlorophyta)	
Caulerpa lentillifera J. Agardh	Philippines Japan
Enteromorpha sp.	
Monostroma nitidum Wittrock	Japan
	Taiwan, Pr. of China
B. Brown Seaweeds (Phaeophyta)	
Ecklonia sp.	Japan
Eisenia sp.	Japan
Heterochoradaria sp.	Japan
Hizikia sp.	Japan
	Korea, Republic of
Laminaria japonica Areschoug	Japan
L. japonica	China
	Korea, Republic of
Macrocystis sp.	Japan
Nemacystus sp.	Japan
Nereocystis sp.	Japan
Sargassum sp.	Japan
Undaria pinnatifida (Harvey) Sur.	Japan
	China
	Korea, Republic of
U. Peterseniana (Kjellman) Okamura	Japan
U. undariodies (Yendo) Okamura	Japan
C. Bed Seaweeds (Rhodophyta)	
Eucheuma alvarezii Doty	Philippines
E. denticulatum (Burman) Collins et Harvey	Philippines
E. gelatinae (Esper) J. Agardh	China
Gelidium amansii Lamouroux	Japan
Gloiopeltis sp.	Japan
Gracilaria verrucosa (Hudson) Papenfuss	Taiwan, Pr. of China
	Japan
	China
G. gigas Harvey	Taiwan, Pr. of China
G. lichenoides (L.) Harvey	Taiwan, Pr. of China
	Japan
Porphyra angusta Ueda	Taiwan, Pr. of China
P. dentata Kjellman	Taiwan, Pr. of China
P. haitanensis Chang et Zhang Baofu	China
P. kuniedai Kurogi	Korea, Republic of
P. seriata Kjellman	Korea, Republic of
P. suborbiculata Kjellman	Korea, Republic of
P. tenera Kjellman	Japan
P. yezoensis Ueda	Japan
P. quangdongensis Tseng et T.J. Chang	Korea, Republic of China

The culture of the seaweed Porphyra is believed to have started as early as between 1596 and 1614 in Hiroshima Bay utilising pole and net devices originally installed to catch fish. At present, commercial seaweed culture is limited to five countries in East Asia, viz., Japan and Korea (which both grow mainly Porphyra, Undaria and Laminaria), China (Porphyra and Laminaria), Taiwan (PC) (Gracilaria and Porphyra), and the Philippines (Eucheuma spinosium, E. cottonii and Caulerpa lentillifera). Thirty-one species belonging to 18 genera and three divisions are presently cultured in these five countries, of which only three out of the 31 species are green algae.

In 1988, the estimated world seaweed production for use in the manufacture of carrageenan was nearly 68 000 t of dried seaweeds, of which nearly 66% was supplied by the Philippines and the rest by Indonesia, Chile and Canada. The bulk of the Philippine seaweed production consists of Eucheuma produced mainly in the southern part of the country in reef-protected coastal areas. Caulerpa is also successfully farmed in seawater ponds in Mactan, Cebu (Trono, 1986).

In Taiwan (PC), Gracilaria is cultured in ponds formerly used for milkfish, with Pingtung County alone accounting for 110 ha of the total 400 ha of Gracilaria ponds in Taiwan (PC) in 1974 and producing 1 000 t of dried Gracilaria seaweed. In Japan, indoor facilities are used to obtain buds/seedlings for on-growing at sea. The facilities consist of 70-80 cm deep square or rectangular concrete tanks provided with illumination, a temperature control system, and ventilation. The successful cultivation of seaweeds depends on four important factors:

(i) *Type of Seaweeds Used:* The seaweeds cultured must be healthy and resistant to disease and breakage. They must be able to grow fast and give high yields during harvest. During processing, they must have high amounts of dry matter from which will be extracted high concentrations of carrageenan of high gel strength and viscosity.

(ii) *Ecological Conditions of the Farm:* The farm must be well-sited and fulfill the bio-ecological requirements of the culture species. In general, the presence of a particular seaweed species in an area is a good indicator of the suitability of that site for culture of the species under consideration.

(iii) *Access to Sunlight:* Seaweeds being cultivated need abundant sunlight for photosynthesis. Shading by other seaweeds and plants must be prevented by regular inspection and removal of the unwanted plants.

(iv) *The Seaweed Farmer:* The personality and dedication of the seaweed farmer is an important factor since the farmer must visit the farm regularly and carry out routine inspections. Some of the farmer's chores include shaking off silt and other foreign materials from the seaweeds, repairing broken lines, restoring uprooted stakes, and picking up drifting branches of seaweeds.

Table: *Types of planting material and methods of culture for different seaweeds*

Seaweed Groups/Species	*Country Where Cultivated*	*Type of Planting Material and Methods of Culture*
A. Green Seaweeds (Chlorophyta)		
Caulerpa lentillifera J. Agardh	Philippines	Vegetative propagation by cuttings; pond culture
Enteromorpha sp.	Japan	Naturally produced "seeds" grown on hibi nets in open seas
Monostroma nitidum Wittrock	Japan Taiwan, Pr. of China	Hatchery-reared or naturally produced "seeds" grown on hibi nets in open seas
B. Brown Seaweeds (Phaeophyta)		
Ecklonia sp.	Japan	Natural seeding on improved substrates
Eisenia sp.	Japan	Natural seeding on improved substrates or introduction of mother plants or seedlings
Heterochoradaria sp.	Japan	No information available
Hizikia sp.	Japan Korea, Rep. of	Introduction of fertile plants on natural or artificial substrates; seeding of naturally produced spores or embryos on rocks
Laminaria japonica Areschoug	Japan	Hatchery produced "seeds"; rope cultivation in open waters using artificial support system; natural recruitment on improved substrates; stone planting or bottom culture using artificially seeded stones
L. japonica	China	Hatchery produced "seeds"; rope cultivation in open waters using artificial support system; scone planting or bottom culture using artificially seeded stones
	Korea, Rep. of	No information (probably same used In Japan)
Macrocystis sp.	Japan	Natural "seeds" on improved substrates; hatchery produced seedlings on twines introduced to artificial substrates
Nemacystus sp.	Japan	No detailed information available
Nereocystis sp.	Japan	No detailed information available
Sargassum sp.	Japan	Introduction of mother plants or seedlings; artificial substrates in open seas
Undaria pinnatifida (Harvey) Sur.	Japan China Korea, Rep. of	Hatchery produced "seeds"; raft or floating rope system in open seas; stone planting using artificially seeded stones; bottom planting in open seas; management of natural stocks by improvement of substrates for natural seeding
U. peterseniana (Kjellman) Okamura	Japan	Same as used for U. pinnatifida
U. undariodies (Yendo) Okamura	Japan	Same as used for U. pinnatifida
C. Red Seaweeds (Rhodphyta)		
Eucheuma, alvarezii Doty	Philippines	Vegetative cuttings using artificial support system on open reefs
E. denticulatum (Burman) Collins et Harvey	Philippines	Same as used for E. alvarezii
E. gelatinae (Esper) J.	China	Vegetative cuttings tied to pieces of corals and

Trono and Ganzon-Fortes (1988) listed the following criteria for selecting good sites for Eucheuma in open waters and Caulerpa and Gracilaria in seawater ponds:

Unpolluted seawater supply.

(ii) Salinity of 30-35 ppt Eucheuma and Caulerpa and 8-25 ppt for Gracilaria.

(iii) Water temperature of 27-30* C.

(iv) Moderate water movement of 20-50 m/min.

(v) Water depth of 0.5-1 m at low tides and not more than 2-3 m at high tides, and

(vi) Firm bottom protected from strong waves for Eucheuma and muddy-loam bottom for Caulerpa ponds.

Seaweeds are grown using different types of planting material (vegetative cuttings, natural seeds, hatchery-reared seeds) and methods of culture (store planting, bottom culture, rope method, rope-concrete method, and pond culture either in monoculture or polyculture with milkfish, shrimp and crabs). These methods are described in detail by Trono (1986).

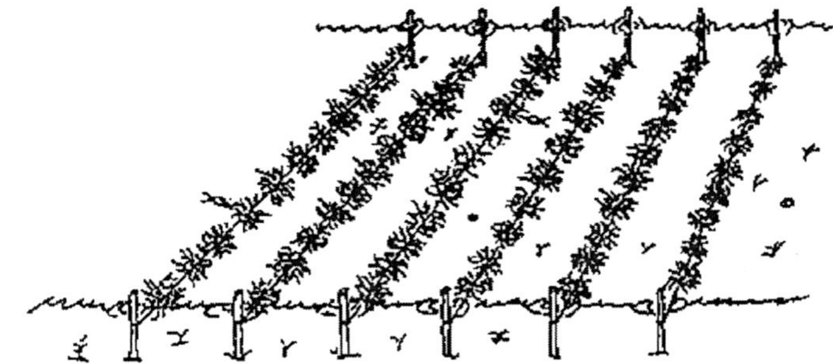

Figure: *Three methods of Eucheuma culture (Monoline Method)*

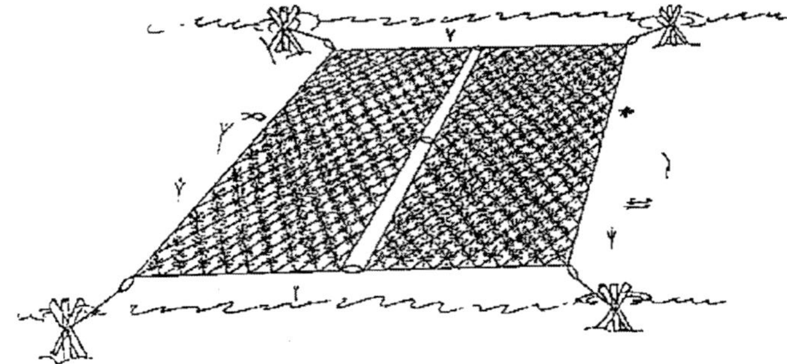

Figure: *Three methods of Eucheuma culture (Net Method)*

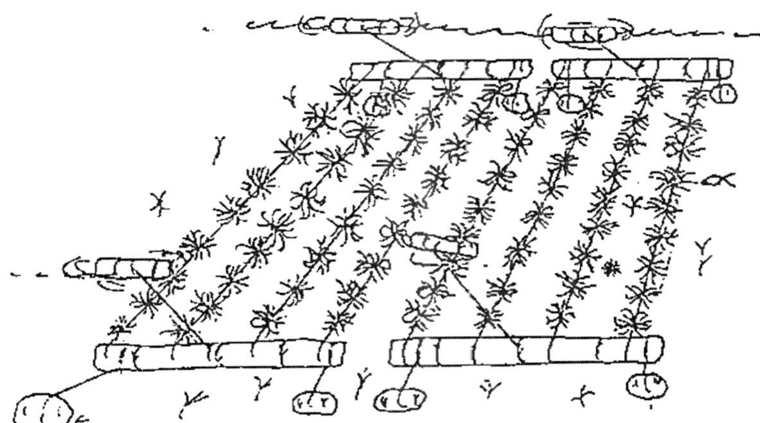

Figure: *Three methods of Eucheuma culture (Floating Method)*

In the Philippines, the monoline method of culture is the most popular and successfully used of these methods (Alih, 1989). The farming activities involved in monoline culture of Eucheuma species based on the Philippine experience are as follows (Trono and Ganzon-Fortes, 1988):

(i) Securing a license from the Bureau of Fisheries and Aquatic Resources (BFAR) prior to farming the area.

(ii) Preparing required materials needed for farm construction.

(iii) Clearing the area of sea grass, seaweeds, large stones and corals, and other foreign materials, followed by measuring it according to the proposed dimensions of the farm. Wooden stakes are then driven into the bottom with the help of an iron bar and sledgehammer and arranged into 10 m rows at 1 m intervals. An 11 m nylon line is securely tied to one end of each stake about 0.5 m above the ground and then stretched to the corresponding opposite stake and tied securely. If the current is very strong, an additional row of stakes is placed in the middle to provide additional support.

(iv) Obtaining seedlings from the nearest source and transporting them to the farm site within the shortest possible time. During transport, the seedlings are protected from exposure to sun, wind, heat or rain. If the transport of seaweeds will take several hours, the seaweeds are kept damp during the trip and upon arrival at the farm, are immediately submerged in water.

(v) Preparing the seedlings by tying bunches weighing about 50-100 g with soft 25 cm long plastic straw, and then tying these to monolines in the water at 20-25 cm intervals. The plants are allowed to grow to about 1 kg or larger before harvesting.

(vi) Building a farm house if drying of the harvested seaweeds is part of the operations. The farm house is built in or near the farm site so as not to waste time during post-harvest handling. The size of the farm house, which is designed to provide for drying and storage, will depend on the farmer's financial capacity and market commitments.

(vii) Maintaining planted seaweeds by inspecting them regularly while they are growing. Unwanted seaweeds which will compete with the Eucheuma for nutrients and sunlight are removed along with dirt and other foreign materials clinging to the seaweeds. Lost or broken Eucheuma are replaced.

(viii) Harvesting the whole plant and reserving select portions as seedlings for the next crop.

(ix) Sun-drying of the rest of the harvest by spreading these on a drying platform of bamboo slots initially lined with coarse fine-mesh nylon net. The seaweeds are freed of all foreign matter clinging to them.

During hot and sunny weather, it takes about 3-4 days to dry the seaweeds to a moisture content of about 30% or less. The dried materials are then packed in plastic sacks for storage in a dry place or for delivery to the buyer.

The pond culture of Caulerpa involves the following major steps (Trono, 1988):

(i) *Pond Construction:* The pond is divided into manageable units measuring about 0.10-0.25 ha. The pond design allows for a flow-through system by providing each unit with its own supply and drainage gates.

Water flows uniformly from the main gate to the secondary and exit gates during the draining and flooding process. Peripheral diversion dikes or canals along the landward edge of the pond are also built to divert run-off water from the ponds during the rainy season.

(ii) *Planting:* To facilitate planting activities, pond water is drained to a depth of about 0.3 m. Caulerpa seedlings are obtained from the nearest source available and transported to the farm site within the shortest possible time.

The ponds are stocked at a rate of 1 000 kg seedlings/ha or 100 g/m^2. A handful of seedlings is uniformly buried on one end at approximately 1 m intervals using a string as guide.

After planting, the pond water is gradually raised to a depth of 0.5-0.8 m or just until the plants can be seen from the surface of the water.

The newly planted seaweeds are inspected after a few days. Uprooted seaweeds are replaced and bare areas are replanted.

(iii) *Pond Management:* Water is changed daily or every other day to maintain adequate levels of nutrients. During the initial stages of growth, the seaweeds deplete the water of nutrients at a high rate and frequent water changes are needed to replenish lost nutrients and eliminate the need to fertilize. Water level is, however, carefully maintained to prevent the collapse of the dikes.

Unwanted seaweeds, sea grasses, and animals which will compete with the Caulerpa for nutrients are regularly weeded out. The dikes and pond gates are inspected regularly to check for leakages, which are repaired immediately. This is vital, especially during the typhoon season.

The application of fertilizer may not be necessary as long as frequent water change is maintained. However, fertilization is resorted to when the stocks appear unhealthy and pale in colour, i.e., from light green to yellowish. When this happens, pond water is changed and fertilizer with a high nitrogen content is applied at the rate of 16 kg/ha by broadcasting or by suspending the fertilizer contained in several layers of plastic sack in strategic areas in the pond. The pond water is not changed in the next two to three days.

(iv) *Harvesting:* Two months after planting, the Caulerpa forms a uniform carpet on the pond bottom, a good indicator for harvest time.

About 75% of the crop is harvested by uprooting the Caulerpa from the mud and placing it on to a wooden raft.

About 25% of the original crop is left behind, uniformly spaced on the pond bottom to serve as seedstock for the next crop. This may be harvested after two to three weeks.

Harvested seaweeds are washed in clean sea water to remove mud and other dirt. The clean seaweeds are then placed in a basket or clean plastic sheets for further sorting and cleaning before packaging and immediate transport to the market.

Selective Breeding - The Key to Sustainability in Aquaculture

Selective breeding is an important aspect of fish farming activities in Iceland. Genetically selected stocks will play an important role in development and ensure the best use of the environment and at the same time improve production cost-efficiency. In Iceland, selective breeding programmes exist for Atlantic salmon, Arctic char and Atlantic cod. The main objectives of selective breeding programmes today are:

- Faster growth
- Later sexual maturity
- Higher resistance to diseases (higher survival)
- Better flesh quality (lower fat-content, colour, texture etc.)

Atlantic Salmon (Salmo Salar)

The Atlantic salmon strain, which Stofnfiskur has been farming since 1991, is called Saga, a stock based on three Norwegian strains, Mowi, Bolak and Sunndalsøra. These formed the basis of the company's early breed selections. Fish produced from the Saga broodstock have proved to be significantly superior to those bred from the originally imported stocks. Compared with other strains, Saga progeny achieve faster growth rates, later sexual maturity and higher resistance to disease. Stofnfiskur is the only company in the world with the ability to produce salmon eggs on a year-round basis, an achievement based on dynamic breed selection and top quality research.

Artic Char (Salvelinus Alpinus)

The Hólar University College Arctic char breeding programme began in 1992, based on a few stocks from the rivers and lakes in Iceland. In 1998, Hólar University College and the Ministry of Agriculture concluded a special agreement pertaining to the breeding of Icelandic Arctic char. The agreement guarantees the breeding project operational capital and sets an administrative framework for it.

Stofnfiskur's Arctic char breeding programme began in 1991, based on stock from the River Grenilækur. The Grenilækur strain is renowned for its dark dorsal side and silvery abdomen. More recently, breed selection activities have been expanded to include two additional strains.

Atlantic Cod (Gadus Morhua)

The company Icecod was established in the year 2003 to take care of the cod breeding program in Iceland. Icecod's main shareholders are Stofnfiskur, Marine Research Institute, Fiskey, Hradfrystihusid-Gunnvor and HB Grandi. Juvenile production is located at Hafnir, Reykjanes peninsula. On-growing is located at the Hradfrystihusid-Gunnvor sea cage station in Isafjardardjup and the HB Grandi sea cage station in Berufjordur. Over 705 family groups were produced from 10 different spawning sites around Iceland. Approximately 350 viable families have been established and will be used to form the base population for the selection programme. The project has a support grant from the AVS R&D Fund of Ministry of Fisheries in Iceland; (AVS stands for Added Value in Seafood).

2

Application of Crossbreeding in Fish and Shellfish

The first step in determining if crossbreeding has a place in the breeding strategy for a particular species is to evaluate all possible crosses between different strains or species for the economic traits in question. If the number of strains available is large, one must select the crosses that are most likely to give valuable results. It may be advantageous to make use of strains with very different origin and also to use strains that, in combination, have favourable traits. In Israel crossbreeding programs are currently in operation, which use strain crosses of common carp. Secondly inbred lines should be developed and the crosses tested under natural conditions to find the most valuable cross for farming. This breeding system aims particularly at utilising non-additive genetic variances. One of the practical difficulties here would be to keep the inbred lines running because of high mortality (inbreeding depression). Bakos (1979; 1987) reported results where inbred lines of common carp were used in a crossbreeding program. Third and last, if possible a reciprocal recurrent selection (RRS) program should be evaluated, to determine the relative importance of general and specific combining abilities. RRS can only be used for multiple spawners and can therefore not be applied in for example Pacific salmon. It would also be very difficult to practice in Atlantic salmon since the majority of males die after the first spawning together with a high number of females. Other species, such as tilapia and rainbow trout, may be more suitable candidates for the application of RRS schemes. One significant advantage, using crossing between selected lines in a breeding program is that this enables the breeders to protect their genetic improved material. By selling crossbred animals only, the pure-bred seedstock are not released.

Crossbreeding Experiments

The status of hybridisation between species of salmonids. He concluded that in most cases the hybrids farmed in the same environment, as the parental species show intermediary or, at best, equal growth to that of the better of the parents. This is in accordance with results from Refstie (1983a) after hybridising of four salmonid species, Atlantic salmon, brown trout, sea trout and Arctic char. Neither growth nor survival of hybrids did exceed the performance results of Atlantic salmon. In some experiments, better results have been found for survival rate, the hybrid often being similar or even superior to the hardiest species of hybrid combinations. Benzie et al., (1995) found no indication of hybrid vigour in growths rates when crossing two species of tiger shrimp, *Penaeus monodon* and *P. esculentus*. Growth rates of the hybrids were similar or lower to that of pure *P.monodon*. Gjerde and Refstie (1984) investigated the heterosis effect between crosses of five Norwegian strains of Atlantic salmon. They did not find a significant heterosis effect for either growth rate or survival rate. Likewise, Friars et al. (1979) found no heterosis effect for growth rate of Atlantic salmon fry. However, in rainbow trout Gall (1975) and Ayles and Baker (1983) reported significant heterosis for body weight among crosses of rainbow trout strains.

Systematic crossbreeding has been performed among common carp varieties. The significance of heterosis for growth rate, survival and cold tolerance among crosses of wild and domesticated European, Russian, Chinese and Japanese strains have been repeatedly reported (reviewed by Hulata, 1995). Wohlfarth (1993) summarised experimental data collected over more than two decades of research in Israel on common carp and concluded "heterosis for growth is a common but not universal phenomenon in carp." As a rule, heterosis was not found when one of the parental lines was Dor-70, a line generated in a long-term mass selection experiment for faster growth. Gjerde et al. (1999) estimated heterosis for body weight and survival in rohu carp (*Labeo rohita*) and concluded that cross-breeding of Indian stocks of rohu carp seems to have little practical significance.

The GIFT crossbreeding experiment with tilapia (Bentsen et al., 1998), already mentioned, showed that only seven out of the 22 crosses that expressed a significant heterosis were better performing than the best pure strain and the largest gain was about 11 %. In general non-additive genetic effects on body weight were modest compared to the additive plus reciprocal effects. Both Wohlfarth (1993) and Bentsen et al. (1998) have reported results that indicate that the expression of non-additive genetic effects may be more sensitive to environmental variation than additive effects. Heterosis may then be poorly expressed in some farm environments because of genotype by environment interactions affecting the nonadditive genetic performance.

In the case specialised hybrids may have to be produced for certain farm environments. Knibb et al. (1997) found that crossbreeding among seabream strains yielded little heterosis, this was explained by lack of inbreeding and genetic differentiation. Reviewing results from several crossbreeding experiments Knibb (2000) concluded that over all species, hybrids usually resemble the average of their parents. Given the large number of attempts to produce new fish hybrids, remarkably few (apparently less than 1 %) have resulted in sustained commercial production.

The search for infertile hybrids, which do not divert food into gonads and therefore have superior production traits, may become very important. As unisexual off-spring have been obtained in several interspecific crosses in tilapia such monosex (male) cultures are recognised as the best solution to over-population caused by the high fecundity of tilapia under almost any pond condition. In addition, males also grow faster than females. Pruginin et al. (1975) list several crosses which have given 100 percent male offspring while Hulata et al. (1983) propose to use progeny testing to ensure getting 100 percent male progeny in tilapia. A promising cross between *Oreochromis niloticus* and *O. aureus* tilapia in Israel is reported to produce nearly all male offspring.

Purebreeding

The breeding method or strategy for additive genetic improvement within a population is known as purebreeding, and the method to choose for continuous genetic improvement over a long period of time. Mating of close relatives must be avoided and individuals that possess a majority of positive (desirable) genes are selected as parents for the next generation. Individuals that possess a majority of positive gene alleles normally show good production results. These "good genes" and properties are transferred to their offspring. Individuals that possess a majority of positive genes are said to have a high breeding value. The breeding value of an individual cannot be measured directly.

Neither can it be measured with 100% accuracy. The true breeding value will therefore remain unknown and to a greater and lesser extent be masked by systematic and stochastic environmental effects and also by effects caused by interactions among the genes.

The breeding value can primarily be estimated by recording the product of the genes, i.e. the phenotypic values of the trait(s) (or the use of genetic markers linked to QTL as described). The phenotypic records may be obtained from the individual itself or from relatives as full- and half-sibs, progeny or parents. Records on relatives can be used because the individual and its relatives share common genes. Information from close relatives is in general more valuable than information from distant relatives. Records on full-sibs are thus more valuable than records on half-sibs because the individual shares a larger proportion of common genes with its full-sibs than with its half-sibs. Records on progeny are of particular interest as the breeding value of an individual is strictly defined as the value of an individual judged by the mean value if its progeny.

Selection Methods

The aim of selection is to identify and select as parents for the next generation the individuals whose progeny, as a group, have the highest possible additive genetic merit for the trait or traits in question. This is almost equivalent to selecting as parents those whose own additive genetic merit is highest. The basic effect of selection is to change the gene frequencies, and the effects that can be observed are changes of the population mean. If there are no differences of fertility among the selected parental individuals or of viability among their progeny, then the gene frequencies are the same in the offspring generation as in the selected parents.

Available selection methods to fish breeders include individual selection, pedigree selection, family selection, within family selection, combined family and within family selection, and progeny testing. The efficiency of each method can be predicted by calculating the expected genetic response for a given set of parameters. The efficiency of selection is partly dependent on how accurately the breeding values of individual animals are evaluated.

For each method, we describe from which individuals or relatives we obtain our information for the estimation of breeding values of selection candidates. The objective for all methods is to maximise the probability of correctly ranking of the potential breeding animals (on true additive genetic merit). This is equivalent to maximising the correlation between the true and estimated breeding value. This correlation (rTI) is frequently termed the accuracy of the breeding values and is an important parameter as it is directly proportional to the expected response to selection. What selection method to choose is dependent on several factors including the heritability of the trait(s), the nature of the trait, recording methods and the reproductive capacity of the species. A description of each method is given in the following.

Individual Selection

Selection based on an individual's own performance or phenotype is called individual or mass selection. This is a well-known and widely used method of selection in animal breeding, and for most aquaculture species the only method practised. Individual selection is usually the simplest method to operate and in many circumstances it yields the most rapid response. However mass selection may be unsuitable if there are large uncontrolled systematic environmental variation (for example age differences). Additionally, there is no control of inbreeding with mass selection, and this has caused problems in a number of fish breeding programs.

A prerequisite for using individual selection is that the trait(s) can be measured on the breeding individual itself while being alive. The method is consequently difficult to practice for meat quality traits, disease resistance and for age at sexual maturity when the frequency of maturing or immature fish is low (<10-15%) and it is not efficient on traits with low heritability. The accuracy of the selection method is equal to the square root of the heritability of the trait (when h2=0.25 accuracy rTI = 0.50).

In fish individual selection is of particular interest for growth rate as the heritability for this trait is fairly high (h2=0.20-0.40) in most species. Since there are many advantages with individual selection, methods should be developed for recording also other traits than growth rate like for example ultrasound of live fish to assess fat coverage. When applying individual selection it is of vital importance that the environmental influence is kept the same for all individuals that are to be compared at any stage of the life cycle. In this manner, the probability of correct ranking of the individuals is maximised. Differences between individuals or groups of individuals for environmental factors like water temperature and salinity, different tanks, ponds or cages, density, light condition, and type of food and feeding regimes, may reduce the accuracy of the selection substantially and in that way reduce the possibility for genetic improvement.

To obtain as equal environmental conditions as possible all individuals to be compared should be hatched at the same day or within a few days period and thereafter reared under identical environmental conditions.

If we want to select individuals reared in different environments, for example in different tanks, or in different cages on the same or different farms; a particular precaution is needed. Due to environmental differences, absolute trait values are not directly comparable. Instead, the mean value for each environmental group should be calculated and the individuals ranked and selected based on their deviations from their group mean.

If the environmental groups are significantly different in their mean values, selection based on deviations will result in more individuals being selected from the group having the highest mean value. This bias is a result of the dependency between the mean and the variance for many traits and frequently observable for body weight. Comparable deviations can be obtained by first multiplying all observations in an environmental group with a constant that is equal to the ratio between the mean value for the group chosen as the base and the mean value for the actual group.

When individual selection is the only selection method practised, it is not usual to tag individuals and in that way keep track of the genetic relationship among them. Neither is it possible from a practical point of view to keep many full and half-sib groups separate until time of selection. Therefore, to avoid mating of close relatives and rapid accumulation of inbreeding, the number of breeding animals should be kept fairly high, at least 50 mating pairs per generation (Bentsen and Olesen, 2002) thereby securing a large effective population size.

Pedigree Selection

By this selection method, breeding animals are selected based on their parents, grandparents or more remote ancestor's performances or breeding values. However, an individual receives a random sample of half of its chromosomes or genes from each parent, opening a vast number of possible new chromosomes or gene combinations among the offspring. This segregation of genes for each new generation may result in substantial deviation in breeding values among offspring.

Selection based on performance of ancestors is generally of limited value if other information is available. The weakness of pedigree information in improving the accuracy of selection lies in the fact that information is generally available on only a few ancestors and the effects of genetic segregation become large after one or two generations.

The accuracy of this selection method will therefore not be high. As a result pedigree selection is not much used as the only method of selection in modern animal breeding.

Family Selection

Family selection refers to a selection method in which family groups are ranked according to the mean performance of each family and whole families are saved or discarded (Lush, 1947). The individuals saved as breeders for the next generation are either all the individuals in selected families or randomly chosen individuals taken equally from all selected families. Because selection is among whole families, the selection differential is a function of differences among families, not differences among individuals.

Individual values are not used except in so far as they determine the family mean. The families may be of full-sibs or half-sibs, families of more remote relationships being of little practical significance.

The efficiency of family selection rests on the fact that the environmental deviations of the individuals tend to cancel each other out in the mean value of the family. Accordingly the phenotypic mean of the family comes close to being a measure of its genotypic mean. The advantage of family selection over other types of selection is greater when environmental deviations constitute a large part of the phenotypic variance. The chief circumstance under which family selection is to be preferred is when the trait selected has a low heritability. With low heritability the use of family mean give an increased accuracy when estimating breeding value. On the other hand, environmental variation common to members of a family impairs the efficiency of family selection. If this component is large it will tend to swamp the genetic differences between families and family selection will be correspondingly ineffective.

To reduce the common environmental component to a minimum, the environment for all families should be standardised as far as possible in the period the families are kept separate. In addition, individuals from all families should be tagged as early as possible and thereafter reared together in the same tank, pond or cage, communal rearing. Refstie and Steine (1978) estimated the common environmental component in Atlantic salmon, due to rearing families in separate tanks for a period of 6 months, and found that this component contributed about 10 % to the total variation in growth rate. If the common environmental component is large, splitting of families in replicate/triplicate tanks should be considered.

Another important factor affecting the efficiency of family selection is the number of individuals in the families; the family size. The larger the family, the closer is the correspondence between mean phenotypic value and the mean genotypic value. The high reproductive capacity in aquatic animals makes family selection important for these species.

Summarising to this point, the conditions that favour family selection compared to individual selection are low heritability, little variation due to common environment and large families.

Another great advantage with family selection is that, based on phenotypic observations on full-sib and/or half-sibs, breeding values can be estimated for traits that cannot be measured on the individuals that are to be used as parents. Carcass quality traits and disease resistance can therefore be included in the breeding objective by applying family selection. Family selection is also far more effective than individual selection for threshold traits such as age at sexual maturity, particularly at low frequencies or incidence of the trait. An incidence of 50 per cent gives the best discrimination between families.

In order to keep the rate of inbreeding low and the intensity of selection high, the number of family groups bred and measured should not be smaller than 50-100. In the period prior to tagging, the family groups have to be kept separately. This makes family selection costly in terms of space. If breeding space is limited in this period, the intensity of selection that can be achieved under family selection may be quite small. The accuracy of family selection is dependent on several parameters; the heritability of the trait (h2), the family size (n) and family type (full and/or half-sibs), and the variation due to common environment (c2).

With a family structure of full-sibs only, the upper limit of the accuracy is rTI = 0.71, which is the square root of the additive genetic relationship among full-sibs, assuming the environmental variance common to full-sibs is equal to zero. There are two major limitations with family selection. Intensive family selection can quickly result in rapid accumulation of inbreeding because whole families are selected. Another weak point of family selection is that as only 50 % of the additive genetic variation is expressed between families, only 50 % of the variation can be utilised. The other 50 % of the additive genetic variation are expressed within families.

Prediction/Expected Response

Selection response, measured as incremental improvement per year, is the most critical aspect of the efficiency of a breeding plan. The length of time required to complete a generation, the generation interval, can be excessive for some selection methods. When this occurs, a good response per generation may not be practical due to the number of years required to achieve the response. It is also difficult to discuss selection methods because the procedures involved can have a major effect on selection intensity. To overcome this difficulty, selection responses for various selection methods are standardised by calculating responses as a ratio using responses to individual selection as a common denominator. One other practical consideration is the influence of various selection methods on rate of inbreeding.

Genetic gain depends on:

- how well the animals are evaluated, or the accuracy of prediction
- the amount of selection or selection intensity
- the magnitude of the genetic differences among animals or standard deviation of additive genetic values and
- how rapidly better younger animals replace their parents, known as generation Interval

Correlated Response

Natural selection will enable animals to adjust to their environmental conditions whether they are in the wild or under crowded farming conditions. Natural selection is therefore important for domestication. Artificial selection for economic important traits such as growth rate and survival will facilitate the rate of domestication (Doyle, 1983). The result will probably be animals that adapt better to life in captivity with lower levels of stress. Farmed Atlantic salmon in Norway have undergone seven generations of selection for increased growth rate. The fish in the later generations seems to be less sensitive to environmental stress than genetically wild fish. These changes are difficult to measure, but are easily seen.

Correlated response to artificial selection depends on the genetic correlation between the traits, their heritability, the selection differential and the phenotypic variance for the trait not directly selected for. Two causes of genetic correlation between characters are described in breeding theory; linkage and pleiotropy. Linkage is the situation in which different loci are situated close together on the same chromosome preventing the genes from segregating independently at meiosis. Pleiotropy is simply the situation where a single gene affects two or more traits. The actual genetic correlation between two traits is the net effect of pleiotropy and linkage.

In livestock species negative side effects of selection for high production efficiency seems to be more at risk for behavioural, physiological and immunological problems (reviewed by Rauw et al., 1998). Beilharz et al. (1993) explained these side effects by the resource allocation theory, which states that fitness is a product of many major fitness components. Consequently when planning and implementing selective breeding in aquatic species precautions to avoid negative side effects of selection should be taken. Careful monitoring of possible correlated traits is needed and more basic knowledge of welfare and behavioural needs.

Indirect Selection

In a selection program it may be desirable to include correlated traits with no economic value in order to improve a trait through its correlated response rather than select for it directly (Gjedrem, 1967). If we want to improve character A, we might select for another character B, and achieve progress through the correlated response of character A. This is indirect selection; selection applied to some character other than the one it is desired to improve. Indirect selection can only be expected to be better than direct selection if the correlated trait has a substantially higher heritability and the genetic correlation is high (Falconer and Mackay, 1996). However, practical considerations may make indirect selection preferable.

There are three practical considerations that may make indirect selection preferable:

- If the desired character is difficult to measure with precision, the errors of measurement may so reduce the heritability that indirect selection becomes advantageous. This is frequently true for disease resistance
- If the desired character is measurable in one sex only, but the secondary character is measurable in both, a higher intensity of selection will be possible by indirect selection
- The desired character may be costly to measure as for example the efficiency of feed conversion. Then it may be economically better to select for an easily measured correlated character, such as growth rate.

In fish disease resistance is a candidate trait for indirect selection since disease resistance is difficult to measure on individuals. Today this can be done by including survival results after challenge tests in the breeding goal or survival results during grow out. One major drawback for both methods is that it can be done on a family basis only. By using an indirect measure correlated to disease resistance this could be done on an individual basis. Immune parameters and physiological parameters such as lysozyme (Røed et al., 1992; 1993) haemolytic activity (Røed et al., 1990; 1992; 1993) antibody response to specific antigens (Eide et al., 1994; Strømsheim et al., 1994a, b; Lund et al., 1995; Fjalestad et al., 1996), ÀÛDÜ2makroglobulin, antiplasmin (Salte et al., 1993) and cortisol (Refstie, 1982; Fevolden et al., 1994) have shown genetic variation and are examples of possible traits for selection for increased disease resistance in Atlantic salmon. They can be all recorded on the breeding candidates and their full and half-sibs. Since blood samples can be taken prior to selection of broodstock the generation interval will not be increased.

Indirect selection for survival based on a correlated trait with no economic value results in a much lower response in survival compared to direct selection. This is the expected correlated response in survival when information of a correlated trait is included in the selection index and survival is the only trait included in breeding goal. The genetic correlation between survival and the correlated trait must be large if indirect selection shall compete with direct selection. The same would apply to low survival rates.

By combining direct and indirect selection for survival (at an overall survival of 50 %) the expected response increases by 5 %

(family selection) or 8 % (combined family and individual selection) when the correlated trait has a heritability of 0.3 and a genetic correlation of 0.3 to survival.

Design of Breeding Programs

The prospects of substantial genetic improvement have been well documented in several fish species (Gjedrem, 1997b). This potential should be exploited through effective and sustainable genetic improvement programs to develop strains for aquaculture with better performance, resource efficiency and product quality. Well-designed programs are also needed for the sustainable use of genetic resources as a safeguard against future changes in both production and market conditions (Hammond, 1994).

The term design may be used to characterise the size and structure of the breeding nucleus with respect to the number of full-sib families and breeding candidates tested per round of selection as well as the mating and selection strategies applied. An optimum designed program may be defined as one that maximise the genetic gain for a given trait (or the breeding goal) over a given period of time for a set of predefined constraints. Important constraints are the available testing capacity in terms of number of families (tanks) and breeding candidates that can be tested and the tolerable rate of inbreeding. Of importance could also be the establishing the base population, methods to trace pedigree (e.g., physical tags versus DNA tagging) and the degree of connectedness required across levels of fixed effects (e.g., cohorts, test environments, generations) to obtain unbiased estimates of breeding values and genetic gain. The design of the multiplier units is also important, but so far no study on this has been published.

The first large scale selective breeding program for farmed fish was set up for Atlantic salmon in the nineteen seventies (Gjedrem, 1992; Gjøen and Bentsen, 1997). The design of this program was based on basic knowledge in quantitative genetics, experiences from livestock programs and available technologies. Full-sib families were reared separately until tagging size and family identification was obtained by cold-branding and fin-clipping (Gunnes and Refstie, 1980). With the exception of some improvements (i.e. more traits in the breeding objective, more families tested and earlier and individual tagging using passive integrated transponder (PIT) tags) rather few changes have been taken place since then. The establishment of other family based programs for salmonids and other fish species, developed mainly in the late nineteen eighties and nineties, have to a large extent followed the same design.

The possibilities and constraints for the design of fish breeding programs have been discussed in general terms (Gall, 1990; Bentsen and Gjerde, 1994; Gjerde and Rye, 1997; Gjerde et al., 2002), but studies on optimum designs are few and limited to programs where truncation selection has been applied for a single normally distributed trait under individual (mass) (Gjerde et al., 1996; Villanueva et al., 1996; Bentsen and Olesen, 2002) and index selection (Villanueva and Woolliams, 1997; Trong, 2004). The design of breeding programs may be addressed from two different angles (Villanueva et al., 1996). *A posteriori*, the problem refers to methods of selection and evaluation for reducing inbreeding in the course of operation the breeding program. Such methods have received considerable attention in recent years (e.g. Toro and Perez-Enciso, 1990; Villanueva et al., 1994; Caballero et al., 1996; Meuwissen, 1997) and have been efficient in reducing rates of inbreeding with a minimal loss in genetic gain. However, they do not necessarily give maximum gains after a specified number of generations of selection. *A priori*, the problem is to maximum response given the basic design variables: available resources, traits and time scale objectives and attitude to risk (inbreeding). This will determine the value of additional resources, more of different measures or the expected benefits and losses from increased risks (Villanueva et al., 1996). Some of the proposed methods (e.g. different mating designs, non-random mating and optimum contribution selection) can be considered to be applicable to both problems.

The focus in this chapter will be on *a priori* methods and on the opportunities and challenges in designing fish breeding programs taking into account the reproductive characteristics of fish species as well as recent and possible future technological developments that may influence the design of fish breeding programs.

Base Population

The first step to start a breeding program is to collect the genetic material that forms the base population. In Norway, genetic material for Atlantic salmon was sampled from 40 river strains (Gjedrem et al., 1991) and no restriction was placed on strain contribution during the first generations. This resulted in an initial selection between strains. An alternative is to primarily mate animals from different strains before starting selection and apply low selection intensity during the first generations of selection. This strategy was used for the base population of a breeding program for Nile tilapia in the Philippines (Bentsen et al., 1997). This may secure the maintenance of a broad genetic variability (allelic diversity) that would allow long-term selection response and a stepwise inclusion of new traits in the breeding goal. In fish breeding, no study has focused so far on the effect of the design of the base population (i.e. the number of individuals to be sampled from one or several founder strains, their mixing, and the intensity of selection to be applied during the initial generations) on the magnitude and variability of the long-term selection response (risk) and inbreeding. It is worth noting that molecular measures of genetic diversity using neutral markers do not seem to be informative about the magnitude of quantitative genetic variation for traits of economic importance in a population (Reed and Frankham, 2001). Assessment of the magnitude of genetic variation for economic important traits applicable for farmed fish therefore should be obtained from traits recorded on fish raised in a commercial farm environment.

Maintenance of Additive Genetic Variance

Breeding programs may be expected to be in place for many generations, especially for species with a short generation interval. Accumulation of inbreeding over generations reduces fitness in the population (e.g. reduced reproduction, viability and survival), reduces performance (inbreeding depression) of traits directly or indirectly (correlated responses) selected for and reduces the magnitude of the genetic variation thus jeopardising further genetic improvement. Consequently, maximisation of short-term gains does not necessarily imply maximum long-term gains. Inbreeding also influences the variability of response, which determines the predictability of the future genetic mean of the population. The rate of inbreeding is thus an important parameter and has, as will be shown later, a major influence on the design of breeding programs.

The disequilibrium genetic variance (with unlinked loci) is reduced with one-half due to recombination of gametes in the progeny in each subsequent generation. This reduction, however, does not go on indefinitely. A balance is soon reached where the reduction in the disequilibrium variance due to selection is balanced by new variation created by recombination (Mendelian sampling variance). In the absence of linkage three or four generations of selection are sufficient to bring the population near to this balance. In the presence of linkage; the disequilibrium variance is reduced by a fraction less than onehalf in each generation and thus it takes longer to reach the balance. The reduction in the genetic variance due to selection in the initial generations is often termed the Bulmer effect.

The proportion of the additive genetic variance in the base population remaining when the balance has been reached (the stabilised value of the utilisable additive genetic variance) decreases with increasing selection intensity and accuracy of selection. Fimland (1979) shows this for different values of the intensity and accuracies of selection when an infinitesimal genetic model and infinite number of individuals in the breeding population is assumed (i.e., no accumulation of inbreeding). Assuming a finite population and a rate of inbreeding of

F=0.5 % or

F=1% per generation shows the utilisable additive genetic variance over 10 generation of selection for four combinations of the intensity and accuracies of selection. These results illustrate the importance of maintaining the additive genetic variance in the base population by applying moderate selection intensities and rates of inbreeding. Of importance is that a 50 % increase in the accuracy of selection results in a much larger reduction in the genetic variance than a 50 % increase in the selection intensity.

Distinctive reproductive characteristics in fish allow for accurate estimation of breeding values and for high selection intensities and therefore high short-term selection responses. However, high selection intensities and accuracies may result in rapid accumulation of inbreeding and this can jeopardise further genetic improvement, reduce fitness and performance and also produce highly variable responses (Meuwissen and Woolliams, 1994b). Implementation of measures for restricting inbreeding is therefore essential in fish breeding programs.

Strategies to control the rate of inbreeding may include increasing the size of the breeding population, increasing the number of parents selected, limiting the use of each parent, limiting the number of individuals selected from each family and choice of mating design (e.g. factorial versus nested design; Woolliams, 1989). However, in a practical situation such procedures may have to take place for a given set of facilities. For instance, in a situation where full-sib families are reared separately until a size suitable of tagging and for a given mating system, the number of parents selected cannot be increased unless more tanks are made available.

A series of methodologies have been developed for finding optimal designs of breeding programs that give maximum genetic gains with constraints on inbreeding in schemes with discrete generations under individual selection (Gjerde et al., 1996; Villanueva et al., 1996, 2000), index (Villanueva and Woolliams, 1997) and BLUP (Meuwissen, 1997; Grundy et al., 1998) or in schemes with overlapping generations under BLUP selection (Meuwissen and Sonesson, 1998; Grundy et al., 2000). Some of these methods (Meuwissen, 1997; Grundy et al., 1998; Meuwissen and Sonesson, 1998; Grundy et al., 2000) are dynamic, in that they adapt to current selection candidates (i.e., optimum contribution selection), and can therefore correct skewness in contribution of families over generations. Skewness of contributions by families results in higher rates of inbreeding.

The optimisation tools generally deal with normally distributed traits that can be measured on live breeding candidates. Many traits of economic importance in fish do not fulfil these requirements (e.g. survival, carcass quality and disease resistance traits). Thus there is need to investigate and possibly extend the current theory to provide selection tools for scenarios relevant to fish species.

Most current fish breeding programs have closed breeding nuclei. Low levels of migration (i.e., an open breeding nucleus strategy) are effective to break up inbreeding and reintroduce genetic variation (Falconer and MacKay, 1996). The migrants could come from selected populations with similar genetic performance. However, introduction from inferior (wild) populations is often the only option for the introduction of new genetic variation (Osman and Robertson, 1968). In such instances it is difficult to maintain the improved performance of the improved population. New genetic variants also arise infrequently by mutation. The balance between new genetic variation by mutation and loss of genetic variation due to inbreeding will depend on the relative rates of mutation and inbreeding.

Genetic gain is often achieved at increased risk (variance of response). There is a strong relationship between inbreeding and variance of response, such that addressing one of these two components of risk usually suffices and results indicate that large reductions of variance of response and rates of inbreeding can be combined with a small reduction in genetic gain (Meuwissen and Woolliams, 1994a). Nicholas (1989) suggested that to be confident that the expected response will actually be achieved in practice, acceptable coefficients of variation of genetic gain per generation should be not higher than 10% after ten years of selection.

Mating Design

A prerequisite for running a sustainable genetic improvement program is that a reasonably high number of full- and half-sib families can be produced in a controlled and reliable manner. The possibility for separate collection of eggs and semen in many fish species facilities a wide range of mating designs. In some species (e.g. salmonids) a high number of large full-sib and maternal and/or paternal half-sib groups may be produced from a set of simultaneously artificially stripped breeders and spawning may often be synchronised or induced.

For some species artificial stripping is difficult to perform or it may be difficult to obtain a sufficient number of synchronized breeders that can be stripped and mated. For these species the production of full- and half-sib families may be obtained through natural spawning by keeping pairs of fish in separate tanks. This has proven successful in Atlantic cod (Terjesen et al., 2004). The newly fertilized eggs are collected through the outlet of each tank. Paternal or maternal half-sib families were obtained, by first mating one pair of parents, then replacing one of the parents in the tank. Choice of the most appropriate mating design is dependent on several factors. Most important is whether stripping or natural mating can be performed, the availability of sexual mature breeders of each sex and also the type of genetic effects (i.e., additive and non-additive genetic) for which unbiased and accurate parameter estimates or predictions of breeding values are to be obtained. For example, significant non-additive genetic variation for harvest body weight has been detected within fish populations (Rye and Mao, 1998; Pante et al., 2002) and in such a case factorial or mixed nested/factorial may provide more accurate estimates of both additive and non-additive genetic effects than can be obtainable in a purely nested design.

Mass Spawning

Natural mass spawning in which several sexual mature male and female breeders are kept in the same tank is often used when producing grow-out animals, in particular for species where artificial stripping is difficult to perform. However, the relative contribution of each breeder to the total number of offspring in the breeding population will be unknown. If individual selection is to be applied, the selected breeders may be the offspring from a rather limited number of parents. Therefore, the effective population size may become low and consequently the rate of inbreeding high as demonstrated in the gilthead sea bream (*Sparus aurata*) where the effective population size, calculated from the number of offspring sampled from a single day spawning, was less than one third of the actual number of parents used in the mass spawning (Brown et al., 2004). However, the use of DNA markers for parental assignment may overcome this problem. Use of DNA markers for parental assignment in mass spawning design may also facilitate the use of sib records in the selection decisions. However, due to the unknown and most likely very variable number of offspring from each parent, the number of breeding candidates needed to be genotyped may become high (section 12.9). Consequently, natural mass spawning is not recommended in the breeding nucleus.

Single Pairs

This is the simplest mating design in which the milt from each male is used to fertilize the eggs from only one female. Only full-sibs are produced, the number being equal to the number of males or females. In this design additive genetic and other effects common to full-sibs (non-additive genetic effects and maternal and common environmental effects) are confounded and cannot be assessed independently. This may result in biased breeding values with low accuracy, resulting in low genetic gain. Consequently, single pair mating should not be used unless the non-additive genetic effects and the effects common to full-sibs other than additive genetics are low.

Nested

The most commonly used mating design in today's fish breeding programs is the nested design. Milt from one male is used to fertilize eggs from several females (females nested within males). Thus both full- and paternal half-sibs are produced. The number of full-sib groups is equal to the numbers of females and the number of half-sib groups is equal to the number of males. In this design the sire and dam component of variance each account for one quarter of the additive genetic variance. However, the dam component may be inflated by non-additive genetic (dominance), maternal (e.g. caused by different egg sizes among the female breeders and possible cytoplasmatic genetic effects) and common environmental effects caused by separate rearing of full-sib families until to tagging.

Alternatively, the eggs from one female may be divided into portions and each portion fertilized with milt from different males (males nested within females). Thus both full- and maternal half-sibs are produced. This design may be a good alternative, particularly in cases where the availability of an adequate number of female breeders is less than that of male breeders. The number of full-sib families is equal to the number of males and the number of half-sib families equal to the number of females. In this design the dam component of variance accounts for a quarter of the additive genetic variance but may be inflated by maternal effects. Consequently, this design will in general provide a less accurate estimate of the additive genetic variance in the population than the sire component estimate derived from the design.

Full Factorial

Semen from each male is used to fertilize eggs from several females and, at the same time eggs from each female are divided into portions and fertilized with semen from different males. Thus both full-sibs as well as paternal and maternal halfsibs are produced. For each set produced the number of full-sib groups is equal to the number of males multiplied with the number of females. The number of paternal halfsib groups is equal to the number of males and the number of maternal half-sib groups is equal to the number of females. This is a good design to obtain reliable estimates of both additive and non-additive genetic variance in the population. However, a serious drawback is that a low number of male and females will be tested for a given number of rearing tanks available.

Partly Factorial

This design may be a good alternative to a nested or factorial design. For a given number of available tanks, a higher number of male and female breeders will be tested than in a factorial design while at the same time obtaining better connectedness between the full- and half-sib families than in a nested design. In the presence of a tank effect, genetic parameter estimates (i.e., heritability) with higher accuracy and precision were obtained from a partly factorial than from a nested design (Berg and Henryon, 1998). Most likely this will also be the case for the estimation cf non-additive genetic Effects.

Future Mating Design

Use of optimum contribution selection (e.g., Meuwissen, 1997) implies that not a specific mating design will be used. Rather the output from this selection algorithm will give information about how many offspring (or mates when assuming an equal number of offspring per mating) a given animal should get to maximise the genetic gain while at the same keep the rate of inbreeding at an accepted level. Thus the developments of new selection tools may change the above strict systematic mating design to more mixed designs.

Mating Design and Inbreeding

For a given testing capacity (i.e., number of full-sib families that can be reared separately until tagging) different mating designs allow for different number of sires and dams to be tested. If a given number of scored selection candidates and truncation selection is assumed, the different mating design will result in different effective population sizes and rates of inbreeding. Assuming no selection and random mating. The lowest rate of inbreeding is obtained with the paired mating design. For nested mating designs the rate of inbreeding increases with increasingly skewed sex ratio. The mixed nested/factorial and the pure 2 x 2 factorial designs give the same rate of inbreeding. Factorial designs with each sire (dam) mated to more than two dams (sires) result in fewer sires and dams tested and thus in higher rates of inbreeding. When selection was applied in MOET (Multiple Ovulation and Embryo Transfer) dairy cattle schemes, factorial design resulted in lower rates of inbreeding than nested design without reduction in genetic gain. Studies on appropriate mating design in fish breeding programs are required.

Individual Selection

Optimum designs for individual selection programs in fish breeding under constrained inbreeding through stochastic simulation. For a given number of scored selection candidates (N = 300 to 9600 animals), and mating ratio (d =2, 6 and 10 dams per sire), rates of genetic gain and inbreeding were obtained separately for different number of sires. Designs given a rate of inbreeding close to the specific value were defined as optimum.

As expected the genetic gain increases with increasing population sizes, in particular for small populations and high heritabilities. Imposing a constraint on inbreeding resulted in lower genetic gain, in particular for high heritabilities.

The desired rate of inbreeding has a large influence on the design, in particular for high heritabilities. Thus for the largest population size studied (N = 9600) and for h2 = 0.2 and d = 2, the optimum design uses approximately 35 sires (and 70 dams and 274 animal per full-sib family) for = 1% while 90 sires were used (and 180 dams and 53 animal per full-sib family) for = 0.5%.

The mating ratio has only a minor effect on the genetic gain but a large influence on the design, in particular for high heritabilities. Thus for N = 9600, h2=0.2 and = 1% the optimum design was approximately 45 sires (and 90 dams and 107 animal per full-sib family) for d=2 while it was 25 sires (and 250 dams above-mentioned study (Gjerde et al., 1996), for a given set of value of N, ÀÛûÝF, d and h2, several simulations must be run to find the optimum solution. The number of selected animals is changed in different simulations until the desired level of inbreeding is achieved. This procedure is very computationally demanding, in particular for large population sizes.

Within Family Selection

When selection is practised only for trait(s) that can be recorded on the live breeding candidates, a within-family selection protocol (called 'walk-back' selection, Doyle and Herbinger, 1994) has been proposed. In this protocol, DNA markers are used to identify the pedigree of the potential breeders: First all breeding candidates are graded with respect to the trait(s) selected for. Then the most superior (e.g. the largest) animal is DNA fingerprinted and chosen to become a parent of the next generation. Then the second-largest animal is genotyped and added if it comes from a different family than the first, otherwise it is discarded. The third-largest individual is fingerprinted and accepted if it is not a full-sib of the first two. This process is repeated until a sufficient number of pairs of selected breeders, each from a different full-sib family, are obtained. The protocol maximises the effective population size, but is most likely not to be optimal in terms of genetic gain. Also, with large and variable family sizes the number of fish that need to be genotyped may become very high with a correspondingly high cost. Because the probability of selecting related animals increases with increasing heritability, the number of animals to be genotyped will be higher when selection is practised for trait(s) with high than for trait(s) with low heritability.

Recently the 'walk-back' procedure has been further developed to be applicable also for individual selection (Trong, 2004). If the average coancestry (additive genetic relationship) among the selected candidates (i.e., the inbreeding level in the next generation) is found to be too high, additional fish are selected and their genetic relationship identified. This procedure is repeated until the desired level of relatedness among the breeders is accomplished. This may be further developed to also utilise sib information.

Individual Versus Blup Selection

Gjøen and Gjerde (1998) used stochastic simulation to find optimum design of breeding programs when applying truncation selection on BLUP (BTS) or phenotypic values (PTS). When the constraint on inbreeding was set equal to = 1 % per generation, genetic gain from BTS and PTS was similar for h2 = 0.10, but PTS produced higher genetic gain than BTS for h2 = 0.20 and h2 = 0.40. The reason is the higher selection intensity for PTS more than compensates for the higher accuracy of selection for BTS. The breeding design is quite different for the two selection methods with BTS having substantially large number of full- and half sib families, but the size of each of these families is smaller. Consequently the costs of BTS and PTS schemes may be quite different, in particularly if the full-sib families of the BTS scheme have to be reared separately until a size suitable for tagging.

A special BTS scheme, where costs of rearing the fish until tagging will be the same as for PTS schemes, occurs when candidates are pooled at fertilization or shortly thereafter and DNA markers are used to assign offspring to parents (Sonesson et al., 2004). After tagging, the cost of rearing fish should be the same for BTS and PTS provided the same number of fish are tagged.

Thus it may be argued that comparison of schemes should be done with a constraint on the number of tanks to rear the full-sib families as was done by Sonesson et al. (2004). These authors found that for the same number of selected sires and dams, the rate of inbreeding was much higher for the BTS than for the PTS schemes, while genetic gain was quite similar. At the same level of inbreeding the design of the BTS and PTS scheme was quite different with BTS having substantially large number of fulland half sib families, but the size of each of these families is smaller.

As an aside, the danger of using the simple formula of Wright (1931) can be demonstrated by the results where the rate of inbreeding, as measured by the pedigree, was restricted to be 1 % per generation. If the simple formula of Wright (1931) is used to calculates the rate of inbreeding, the predicted rate of inbreeding was 0.55 % for PTS and 0.18 % for BTS; thus a substantial underestimating of the rate of inbreeding and more so for BTS for PTS.

SIB Selection

When traits cannot be recorded on the live breeding candidates (e.g., carcass quality and disease resistance traits), selection within families cannot be applied and consequently the selection intensity obtained is that between full- and half-sib families (sib selection). The lack of appropriate technologies for recording such traits represent a challenge in fish breeding programs and may have a substantial effect on the design of the programs.

When such technologies become available, the same genetic gain could be obtained with a lower number of selection candidates and families, or higher gains could be obtained with the same number of individuals and families. Studies on the optimum design of fish breeding programs when applying sib selection has not been undertaken. The development of technologies to record such traits should be encouraged.

Progeny Testing

Most fish species, as opposed to salmonids, are multiple spawners making progeny testing of both males and female breeders possible. For male breeders of any species the same could be obtained if reliable procedure for cryopreservation of milt is available. Selection could thus be performed among the parents based on their progeny as well as own and sib records.

The highest ranking could be used to produce a second crop of progeny, either for use in the nucleus or for the production of grow-out animals. Optimum mating design utilising progeny test records will require that each male (female) be mated to more females (males) than when utilising only own and/or sib records.

Selection and Mating

Optimisation tools have mainly dealt with selection decisions, assuming that the selected parents are mated at random and using a given mating design. The reproductive characteristics of fish imply a high potential for non-random mating strategies that have not been investigated to date and few attempts have been made to optimise simultaneously selection and mating decisions. However, a number of non-random mating systems have been proposed to control inbreeding, implying that their application could lead to higher gains without increases in inbreeding. This may include compensatory mating (Caballero et al., 1996), minimum (Toro et al., 1988) and average co-ancestry mating applicable for discrete (Meuwissen, 1997; Kerr et al., 1998; Sonesson and Meuwissen, 2000) as well as overlapping generations (Meuwissen and Sonesson, 1998).

Use of optimum contribution selection (e.g., Meuwissen, 1997; Grundy et al., 1998; Meuwissen and Sonesson, 1998; Grundy et al., 2000) implies that not a specific mating design will be used. Rather the output from this selection algorithm will give information about how many offspring (or mates when assuming an equal number of offspring per mating) a given animal should get to maximise the genetic gain while at the same keep the rate of inbreeding at an accepted level.

For low number of selection candidates (i.e., ÀÛ"Ü 512 newborn candidates per year) optimum contribution selection has given equal or higher genetic gain than BLUP selection and with a substantially lower number of parents selected (Meuwissen, 1997; Meuwissen and Sonesson, 1998). In traditional farm animals the optimum number of offspring may be difficult to obtain because of reproductive limitations, particularly in females. However, in fish species such reproductive limitations are in general less, and therefore optimum contribution selection is of particular interest for fish species. The application of this new selection tool for fish species should be studied and in particular for larger population sizes than done so far. Its influence on the accuracy and unbiasedness of genetic parameters should also be investigated.

Connectedness

Breeding programs need efficient procedures to obtain accurate and unbiased estimates of breeding values to maximise genetic gain and to obtain unbiased estimates of genetic changes. Unbiased estimates can only be obtained if environmental and genetic differences across levels of fixed effects are accounted for (Sorensen and Kennedy, 1984a; 1984b). BLUP methodology offers this possibility provided that there is sufficient connectedness in the data (Kennedy and Trus, 1993; Hanocq et al., 1996). Studies on the level of connectedness needed in fish programs for unbiased estimates of breeding values have not yet been undertaken.

In fish breeding programs the question of appropriate level of connectedness arises when (a) not all families are reared at all test units in order to save costs; (b) the breeding population consists of several cohorts of families produced at different times of the year; (c) genetic material is exchanged between different breeding nuclei to reduce inbreeding and to increase the intensity of selection; and (d) sires and/or dams are reused over generations (e.g. for monitoring genetic changes).

Connected data can be obtained either from parent-offspring relationships (genetic ties) or from parents having offspring with records (direct ties) across different levels of fixed effects (e.g. within and across generations).

In some species it may be possible to create an adequate number of direct ties by keeping randomly bred control groups, or by repeated mating of parental pairs that have been used in the past. Maintaining stable control groups requires substantial resources in terms of testing capacity. Repeated mating requires that both sires and dams are kept alive over generations, or that cryopreserved embryos will be available. This strategy may also give biased estimates of gain unless possible non-additive genetic effects can be accounted for. The alternative is to reuse some sires (here cryopreserved sperm could be used) or dams only.

Physical Tags versus DNA-markers for Parental Assignment

By using DNA (genetic) markers for parental assignment in fish breeding programs, the pooling of families at fertilization or shortly thereafter would be possible. This could eliminate the need for costly multitank facilities and the problem of confounding genetic and common environmental effects (Doyle and Herbinger, 1994). However, the use of genetic markers has some potential dangers primarily as a result of low and variable survival rates among the families from the time of pooling the families to the time of selection of parents. Consequently, the number of fish that need to be genotyped may become very high in order to keep the rate of inbreeding at an acceptable level in programs that apply individual selection, and to obtain records on a sufficient number of animals from all families in programs that utilise sib information in the selection decisions.

The number of DNA markers characterised for commercial species are sufficient for correctly assigning offspring to parents (e.g., Norris et al., 2000; Villanueva et al., 2002) but the technology is still very costly, particularly given the fact that fish need to be retyped each time family assignment is required (e.g. if trait records are obtained at different times). The combined use of DNA markers and physical tags can overcome this problem to some extent. Also the breeding candidates need to be physically tagged so that those to be selected can be easily traced at time of mating.

Nevertheless, there are a few companies that use genetic markers in their breeding programs. Further improvement in genotyping technology may reduce the costs considerably. Therefore it is of interest to evaluate the use of genetic markers for parental assignment in fish breeding programs. Using DNA markers may not only reduce the costs of the programs but also increase the genetic gain. In addition, to remove the need for separate rearing, use of DNA markers provides the opportunity to (1) increase the numbers of individuals and families without additional investment in units for separate rearing (increasing thus the scope for genetic improvement), (2) to compute exact genetic relationships among siblings (Norris et al., 2000; McDonald et al., 2004) and (3) to control rate of inbreeding in programs applying individual selection (Doyle and Herbinger, 1994; Trong, 2004). It would also be possible to include markers linked to loci affecting economically important traits as part of the panel of markers used for parental assignment.

Small Scale Breeding Program

For less important economic species, simple programs that utilise only individual selection primarily for growth should be initiated. These programs may later be extended to more advanced programs that also utilise sib information in the selection decisions if production and thus the economic importance of the species increases. Since individuals are not tagged when applying individual selection, the choice of mating and breeding design structures that gives tolerable rates of inbreeding is of particular importance for such programs. Key issues for such programs are the use of a sufficient number of breeders and restricting the number of sibs tested per family (Gjerde et al., 1996, Bentsen and Olesen, 2002). The inbreeding level in the grow-out animals may be kept very low if breeders from different populations or lines are mated to produce the grow-out animals. In the growout phase the individuals from the lines may be kept in the same rearing unit but need to be identified, for example by fin-clipping. Thus an additional line will not require much extra investment and resources.

To account for possible genotype by environment interaction, a sample of the breeding candidates may be raised at different locations representing different commercial farm environments. Some breeders must then be selected from each of these test environments.

Large Scale Breeding Program

Breeding programs for important economic species should utilise both individual and/or family information in the selection of breeders. The design and logistics of a typical large-scale family based fish breeding program are shown.

Breeding Nucleus

The breeding nucleus typically consists of a large number of full-sib and paternal or maternal half-sib family groups, depending on the mating structure. Generations are usually discrete but in some species the use of light regimes makes it possible to produce cohorts of families at regular intervals throughout the year, leading to overlapping generations. Currently, the full-sib families are kept in separate units/ tanks until a size suitable for tagging by physical methods. A sample of individuals from each full-sib family are tagged and raised as breeding candidates preferably in a commercial farm environment.

The high fecundity of fish allows the use of data from full-sib families that are replicated in facilities outside the breeding nucleus, in the genetic evaluations. The nucleus operations could then focus solely on the production, tagging and distribution of test fingerlings from a large number of families. This would help protect the nucleus against accidents, disease breakdowns and would secure the necessary number of individuals to be able to carry out effective selection (Bentsen and Gjerde, 1994). In this case, the breeding candidates could be raised in any type of environment producing quality gametes. Such a strategy will imply sib selection for all of the traits that are selected, including those that can in theory be recorded on live breeding candidates. But most likely growth would still be recorded on the candidates.

Test Animals

Other samples of tagged individuals from each family are raised as test animals in commercial farm environments either at facilities where destructive carcass quality traits can be recorded or at facilities at which disease challenge tests can be performed. Performance data from different commercial farm environments can be useful for selecting fish that are robust to differences in environmental conditions, provided the magnitude of genotype by environment interaction is marginal and does not lead to severe reranking of families in different environments. Sufficiently connected data could be obtained with only some of the families reared in each test environment, saving thus some costs. Studies are needed to determine the minimum number of farm environments needed for eliminating the risk of producing animals only suited for a particular type of farm environment. This will of course depend on the magnitude of the genotype by environment interaction.

Selection

Based on the data recorded on the test animals and the breeding candidates, animals with high breeding values are selected among the breeding candidates to become the breeders for the next generation. After mating the resulting full-sib families are transferred to separate units in the hatchery and thereafter to separate tanks for growing until tagging size.

Multiplier Units

The improved genetic material (surplus milt and eggs from the selected breeders or from other animals with high breeding values) are transferred as fertilized eggs to one or several multiplier units where the animals are raised to sexual maturity to produce eggs/fry that are sold to fingerling producers. Alternatively, fry/fingerlings are produced and are directly supplied from the selected nucleus breeders to fingerling/smolt producers. The fry/fingerling producers raise the animals to a size suitable for the growout producers.

The high fecundity of fish can be fully utilised when producing grow-out animals since these will be slaughtered and inbreeding is not an issue as it is in the breeding nucleus. This makes the expense of maintaining broodstock low and facilitates the concentration of resources into one or a few multiplier units. The genetic gain obtained in the nucleus may in this way be effectively disseminated throughout the industry with a minimum time lag. How this should be done most efficiently will vary with species and needs further study.

Significant non-additive genetic variation has been detected within fish populations (Rye and Mao, 1998; Pante et al., 2002). So that grow-out animals gain the maximum benefit from these non-additive effects, a systematic exploitation of non-additive genetic variation may be possible by repeated mating of progeny tested broodstock pairs or by development of specialised parent lines through some type of recurrent reciprocal selection.

Currently, all grow-out farmers are offered animals with similar genetic material that represents more or less the average genetic level in the nucleus. The potential for producing a more customised genetic material suited for specific farm environments and market segments may be possible. Grow-out animals could for example be produced from the best nucleus breeders for a single or few traits on request (in contrast with the nucleus where selection is for the overall aggregate genotype), preferably from progeny tested male and female breeders with highly accurate breeding values.

However, the number of top quality breeders from the nucleus and consequently the number of grow-out animals may be limited. If early survival can be improved in the highly prolific marine species, a small number of breeders may still supply a large-scale commercial operation. Another alternative could be to establish separate selection lines outside the breeding nucleus.

Each line could be reproduced using top male breeders from the nucleus with highly accurate breeding values for the desired trait(s) and female breeder from the line using individual selection. Individual selection of both male and female breeders within the line is also a possibility, but would be restricted to traits that can be recorded on the live breeding candidates. No tagging of animals would then be required in the line and several lines could therefore be maintained at a relatively low cost (e.g. by keeping several lines, tagged with fin-clipping, in the same rearing unit). After a few generations of selection the line(s) with highest market potential could be multiplied.

In addition to capitalising on additive and possible non-additive genetic effect at the multiplier level, complementary techniques like molecular genetics, sex and chromosome manipulations (Hulata, 2002) may be added to produce grow-out animals with the best performance. These techniques would need to be accommodated in the framework of the current classical selection programs.

Concluding Remarks

Even if remarkable improvements of additive genetic performance have been obtained in some fish species, the application of livestock selection theory to farmed fish is still in its early stages. The opportunities for continued progress are good, but further investigations are needed so that genetic improvement programs can be developed that are optimal both with respect to genetic gain and costs.

Recent and future developments in molecular genetics present the possibility of using genotype selection for performance traits (either through marker assisted or direct selection on quantitative trait loci or QTL's if this information is available). The optimal use of this information may require modification in the present mating and breeding design. The same applies to present technologies like sex and chromosome manipulations (Hulata, 2002). Methodology for efficiently integrating these technologies into classical breeding programs needs further investigation.

Prediction of Breeding Values

In selective breeding programs the main objective is to move the mean value of traits of economical importance in the desired direction. This change in the population mean due to selection is termed genetic gain. Genetic gain is obtained by selecting as parents for the next generation individuals that have a higher probability than other individuals in the population for transferring favourable values or genes to their offspring. This chapter deals with procedures to detect the best breeding individuals as parents for the next generation.

Parents pass on their genes and not their genotypes to their offspring. New genotypes are being created in each generation, as a result of coming together of new combinations of genes from the parents. Therefore the parents for the next generation should not be selected due to their genotypic value. A new measure of value is needed which will refer to genes and not to genotypes. This will enable us to assign a 'breeding value' to individuals, a value associated with the genes carried by the individual and transmitted to the offspring.

This new measure is the *'average effect of a gene'* which is the mean deviation from the population mean of individuals which received that gene from one parent, the gene received from the other parent having come at random from the population (Falconer and McKay, 1996). The breeding value of an individual for a given trait is equal to the sum of the average effects of the genes, the summation being made over the pair of alleles at each locus affecting the trait. Therefore, the average effects of the parents' genes determine the mean genotypic value of its progeny. Thus the *breeding value* of an individual can be defined as the value of the individual judged by the mean value of its progeny (Falconer and McKay, 1996). The breeding value for an individual, unlike average effect of a gene, can therefore be predicted based on phenotypic values recorded on its offspring. If an individual is mated to a number of individuals taken at random from the population, then its breeding value is twice the mean deviation of the progeny from the population mean. The deviation has to be doubled because the parent in question provides only half the genes in the progeny, the other half coming from the mates selected at random from the population.

Just as the average effect is a property of the gene and the population, so is the breeding value a property of the individual and the population from which its mates are drawn. Therefore when referring to an individual's breeding value, the population in which it was or will be mated must also be specified. Systematic and stochastic environmental effects and non-additive genetic effects will to a greater or lesser extent mask the true breeding value. Consequently, the true breeding value cannot be measured with 100% accuracy.

In animal breeding programs all traits of economic importance should be selected for simultaneously. This is more efficient than selecting for (i) one trait at a time until it is improved to the desired level (tandem selection) or (ii) assigning a threshold value for each trait below which all individuals are discarded, regardless of the superiority or inferiority of their other traits (independent culling).

Different types of methods are available for determining which individuals to be used as parents for the next generation. The methods differ with respect to type of relatives that provide information used for the selection decisions; the breeding candidate itself or relatives like full-sibs, half-sibs, offspring, parents etc. The objective of all these selection methods is to maximise the probability of correctly ranking the animals with respect to their breeding values. For each selection method, different genetic evaluation procedures can be applied to obtain the predicted breeding value. The most commonly used is the Selection Index Procedure (SIP) first applied in plant selection programmes by Smith (1936) and later developed for farm animals by Hazel (1943). Until the early 1970's, this was the standard procedure for the prediction of breeding values in animal breeding.

SIP is described in detail in many animal-breeding textbooks (e.g. Falconer and McKay, 1996; Cameron, 1997) and will therefore only be described briefly here for the sake of completeness. When selection is practised for one trait only and the available information is recorded on the breeding candidate and/or a limited number of relatives, SIP can be illustrated in a very simple manner. In section 13.3 this will be illustrated in detail for different selection methods and combination of methods. An alternative and more powerful procedure is Best Linear Unbiased Prediction (BLUP, Henderson, 1984). In animal breeding this procedure was taken into use in the early 1970's and is now considered the state of the art for genetic evaluation of all animal and plant species. As opposed to SIP, BLUP provides simultaneous estimation of fixed (e.g. environmental) and random animal effects (e.g. breeding values). In addition, additive genetic relationships among all animals in the population can be incorporated and utilised more easily than with SIP.

Best Linear Unbiased Predictor (BLUP)

To assume that all means and variances are known, as required to obtain BLP, is not realistic in practical animal breeding. To do any prediction of random variables, the variances of the population parameters need to be known. Thus, BLUP assumes known variances, but allows the solution of fixed effects (means) to be unknown. In doing so the solutions for fixed effects must be unbiased. This can be obtained by applying a LaGrange Multiplier in deriving the predictor. The appropriate means are obtained as the Generalized Least Squares (GLS) solution for the fixed effect.

In practice, the variances (and the means obtained as the GLS solution) are simultaneously estimated and used as though they were the true values, giving approximate or empirical BLUP (EBLUP). But

BLUP is not necessarily the best predictor among all possible types of predictors, depending on the distribution of the random variables.

Selection Index Procedures (SIP)

SIP provides a reasonable solution to the problem of genetic evaluation of animals in a rather restricted situation; namely, when the animals and traits selected for have the same means or when the first moment (the means) and the second moments (variances) of the distribution of the observations is known. In practical animal breeding this is rarely the case. Therefore, when applying SIP the observed records are adjusted for the actual fixed effects (e.g. age, pond, sex, etc.) and assume that the means thus obtained are the true means.

Genotype – Environment Interaction

Aquaculture takes place under very different environmental conditions. There may be differences in water temperature, salinity, technology or feed for example. No species of fish or shellfish are adapted to all kinds of environments. Outside the environment to which it is adapted, a particular species may not thrive or may not be able to survive at all. For example, salmon cannot compete with tilapia at high water temperatures while tilapia are not able to grow and survive in cold waters. We therefore talk about warm water species, coldwater species and species suitable for temperate climate. Similarly some species are adapted to freshwater, brackish water or salt water. Genotype-environment interaction may partly be ascribed to sensitivity of animals to variation in environmental conditions. Departures in environmental condition from those to which a species is adapted can cause the animal to become stressed. The main factors causing stress-situations for the animals are temperature, salinity, handling and water quality both in running and stagnant waters. Type and availability of feed may also cause interaction. We want robust animals, which will tolerate fluctuations in environmental conditions. In order to avoid developing strains sensitive to different environmental conditions, breeding values should be estimated on data from animals tested under varying farming conditions representative to the industry.

Causes of Interaction

The biological causes of interactions between genotype and environment are many. In short, some genotypes fit exceptionally well in certain environmental niches while poorly into others. The Chinese strain performs better than the other strains in the environments with low growth while it cannot compete with the European and the crossbred strain in the environments with high growth rate. In a statistical sense interactions turn up when the ranking of genotypes are different in different environments. In some data significant genotype-environment interaction may be shown by differences in scale in different environments, such that genotypic differences may be greater in one environment than in the other. This type of interaction can be removed by suitable transformation of data.

A more serious form of interaction is when the best genotype in one environment is not the best in another environment. This causes problems for estimation of breeding values and will reduce the efficiency of selection since one can not be sure that genetic improvement made in a high level of feed and management will be transferred to lower level of environmental conditions. According to Pirchner (1985) genotype-environment interactions are important in situations where environment can not be controlled, such as in plant production, or in animal production under range conditions. In contrast, under farming conditions the environmental factors can be partly standardised and then the genotype-environment interactions are usually of less importance.

Estimation of Interaction

When several genotypes are reared in different environments, an analysis of variance in two way classification of genotypes and environments will yield estimates of the variance between genotypes, the variance between the environments, and the variance attributable to interaction of genotype-environment. A significant interaction need not be a good assessment of the importance of the interaction. The importance of an interaction can be assessed from it's proportion of the total phenotypic variance. If the variation due to genotype-environment interaction account for only a small part of the total phenotypic variation, it will cause minor problems in breeding value estimation. According to Falconer and Mackay (1996) the logical measure for genotype environment interaction is the genetic correlation between performances of genotypes in two environments. A high genetic correlation indicates interactions i.e the same gene affects the trait in both environments. If genetic effects are present in one environment and only partly repeated in another, the genetic correlation will be low.

Solving the Problem of Interaction

The efficiency of a breeding program depends to some extent on the magnitude of genotype-environment interaction. Therefore it is of importance to use some strategies to minimise the interaction. Genotype-environment interaction occurs particularly when the variation in genotypes as well as in environments, is large. When testing families or genetic groups one should avoid large environmental variation. On the other hand one should select environmental conditions for testing families or genetic groups, which are representative for the industry. Already in 1946/47 Haldane and Hammond stated that genotypes would be expressed fully only under a high level of management and recommend that testing of animals for selection should take place in such environments. If the interaction accounts for a large portion of the phenotypic variation or if the correlation between estimated breeding values in different environments is low, the genetic improvement may become small and even negligible. A possible solution to such situation is to develop a strain for each environmental condition to adapt the animals to their demanded environments.

Estimates of Genotype-Environment Interaction in Aquaculture Species

CARP: Twelve different genetic groups of common carp, four strains and their crosses, were reared in five very different pond environments in Israel during 1960 and 1970. Growth rate in the best environment was more than twice as high as in the poorest environment. The presence of genotype-environment interaction was evident from the difference in ranking of the tested groups in the different environments. Especially striking was the difference in ranking of the extreme genotypes between the extreme environments.

In another study, Wohlfarth et al. (1983) compared growth rate of three genotypes of carp, consisting of Chinese and European breed and an inter-racial Chinese x European crossbred. The differences in slopes and intercepts of the response curves of each genotype reared in five different environments demonstrate genotype–environment interaction. This demonstrates that, for the trait growth rate, environmental effects greatly influence the relationship between the two genotypes. In a poor environment the Chinese is dominant over the European, in a good environment the European is dominant over the Chinese, while in intermediate environments the interracial crossbred was overdominant. He found that the genotype-season interaction appear to be of considerable magnitude in common carp and conclude that the measure of genetic variability based upon a single season are unreliable. Further he concluded that genotype-husbandry system interaction is of importance for culture systems using stagnant waters and of less importance in running water. He also pointed out that there might be considerable genotype-competition interaction in polyculture. Genotype-production systems interaction was studied in rohu carp at CIFA (Central Institute of Freshwater Aquaculture), India by Reddy et al. (2001). For the 1993 year class the interaction in body weight between two strains of rohu carp in mono and polyculture was highly significant but accounted for only 0.1 % of the total variation. The interaction in body weight between six strains of rohu in mono and polyculture for body weight was also highly significant in the 1994 year-class, but accounted for a minor portion of the total variation. Similar results were obtained for survival between tagging and harvest.

For the 1994 year-class 57 full-sib groups of rohu were tested in both mono and polyculture. The ranking of the full-sib groups was very similar in these two environments and the genetic correlation between the breeding values in these two environments was high (rG = 0.87) indicating no significant genotype-environment interaction. This result is contrary to Moav (1976). These results strongly indicate that the growth of rohu may be genetically improved through selection because of high heritability (h2 = 0.52) and based on the test results obtained in either monoculture, polyculture or in both, and that genetically improved seed will show its improved growth potential when stocked in either production system.

Salmonides

Broodstock were sampled from 37 rivers/strains for three consecutive years. Samples of tagged smolts from each strain were distributed to five salmon farms along the Norwegian coast where they were reared in sea cages until reaching market size of 5.45 kg after two years. Total data included 36187 tagged fish. The environmental conditions between the farms were very different. The southern farm was located close to the 61 parallel and the northern close to the 67 parallel with differences in water temperature as well as in day length. In each of the three year-classes the growth rate was only half as good in the farm producing the lowest growth as in the farm producing the highest growth rate. Analysis of variance for body weight and length yielded highly significant strain-farm interaction for each year-class. The component of variance of interaction as a percentage of the total variance explained only a small part, between 1.3 and 3.1 %, of the total variance, as shown in Example below. The ranking of strains at the different farms was very similar. It was therefore concluded that genotype environment interaction could be neglected in a breeding program for Atlantic salmon and that only one breeding population should be developed for the country. The data consisted of 24 sires and 119 dam progeny groups reared in 6 different net cages at different sites in Norway. The sire–cage and dam-cage interaction accounted for a large part of the total variation. It was concluded that genotype–environment interaction plays an important role for the trait early sexual maturity. Data of rainbow trout from four year-classes were studied by Gunnes and Gjedrem (1981).

The broodstock originally came from different Norwegian fish farms and was treated as one population. Number of sires/progeny groups was 94 with a yearly variation from 18 to 36 genetic groups. Samples of one-year-old fingerlings from each full-sib group were tagged and distributed to five different farms along the coast. A total of 23086 fish reached 2.6 kg after reared in sea cages for 18 months. The differences between farms were large with about 1 kg difference between the farms with the lowest and the highest body weight of fish. The interaction sire-farm was highly significant but explained only between 1.1 and 5.5 % of the total phenotypic variance for weight and between 0.7 and 4.5 % of the total phenotypic variance for length. Also for rainbow trout it was concluded that genotype-environment interaction could be neglected. An experiment was carried out in Norway and Sweden to investigate genotype environment interaction in farmed rainbow trout. In the experiment 35 males and 131 females were mated hierarchically (Sylven et al., 1991). Individual slaughter weights were recorded on 26663 offspring which were reared in eight representative cage culture systems in Norway and Sweden for approximately 1½ years following the first year in fresh water. The slaughter weight was treated as different traits: 1 in fresh water in Sweden; 2 in brackish water in Sweden; and 3 in salt water in Norway. Variance and covariance components for sires, dams within sires and random errors were estimated. Estimates of genetic correlations were 0.86 for traits 2 and 3, 0.72 for traits 1 and 3, and 0.58 for traits 1 and 2. At least for the latter case some genotype-environment interaction must be assumed.

Tilapia

Eight strains of Nile tilapia were tested for two generations in eleven different environments. The test environments were chosen to cover a wide range of Philippine tilapia farming systems, from simple ponds, to more intensive systems, rice-fish systems and cages. During the first generation, trials using individually tagged fingerlings (total 7652) were communally reared in all test environments for about 90 days. In the second generation trials a total of 3420 fingerlings of pure strains were communally reared in eight test environments (Eknath et al., 1993). The strain-environment interaction was significant but accounted for only 0.3 % of the total phenotypic variation in both generations (Eknath et al., 1993). The results from the first generation are illustrated. In some of the environments with a very low growth rate, there were some differences in rank-order for some of the strains. From Figures it is obvious that the genetic difference between genotypes increases as the environmental conditions measured as growth rate increases.

Catfish

In catfish, Dunham et al. (1990) report that the hybrid between blue catfish and channel catfish grew less than the channel catfish at low densities during the first year of culture in ponds, but that the hybrids grew faster during the second year of culture in ponds. Genotype-environment interaction appeared when the density was increased or if the culture unit was changed to cages. Similarly, some catfish species and their hybrids performed differently in aquaria than in ponds. A notable example was the blue catfish, which competes with the channel catfish in ponds, lakes and rivers, but performs poorly in aquaria.

Shrimp

Oceanic Institute, Hawaii and AKVAFORSK carried out a breeding experiment with *P. vannamei*. Five batches of a total of 294 full-sib groups and close to 18000 tagged progenies were tested in 3 to 4 different farms for growth rate. The family-farm interaction was highly significant for all five batches, but accounted for only 0.5 %, 8 %, 0.1 %, 0.1 % and 0.2 % of the total phenotypic variation for the five batches, respectively. Because genotype-environment interaction was found to be low it is expected that selection based on results from one environment would lead to improved growth in another environment. In another experiment with *Penaeus vannamei,* genotype-environment interaction has been studied by CENAIM (Centro Nacional de Acuicultura e Investigaciones) and AKVAFORSK in

Columbia. 52 full-sib families were tested for growth rate and pond survival in two different farms (Suarez, et al., 1999). A strong positive correlation between family growth performance recorded at two farms (rG = 0.87) strongly indicates that the level of genotype-environment interaction for growth is low. For pond survival the genetic correlation between family means when reared in two different environments was 0.60 indicating the existence of some genotype-environment interaction.

Molluscs

For Pacific oyster Langdon et al. (2000) report a strong genotype–environment interaction in body weight although some families performed well across a range of culture environments. Newkirk (1978) compared growth rate of 4 different populations of *Crassostrea virginica* and their crosses in different salinities and found significant genotype–environment interaction. In scallop Ibarra et al. (1999) did not find significant interaction in growth rate between genotype and tray position. Genotype-environment variation in abalone. Progenies of 100 families were reared in two different farms for 14 months. Genotype–environment interaction was negligible for survival and body size. These examples of genotype-environment interaction have shown that there is considerable difference in the magnitude of interaction from one study to the next, from one species to another and on the extent of the environmental differences. It is therefore not possible to draw general conclusions about the importance of genotype-environment interaction. Therefore for each species, preferably before the breeding program is established, some estimate of the genotype - environment interaction for the traits of economic importance, and for the environments in which the species will be farmed, should be carried out. Usually the genotype–environment interactions increase with increasing genetic distance and increasing environmental differences. As industry increases productivity, and environments become more standardised across farms, it is likely that the magnitude of interaction will be of less importance.

In general breeders should be tested under environmental conditions, which are representative to the industry. However, one should take into account that some results indicate that testing of breeding values should take place under favourable environmental conditions because of larger genetic variation compared with low environmental level.

Significant interaction causing different ranking of animals must be taken seriously when designing breeding plans. When genotype-environment interaction is considerable, a breeding population should be developed for each type of environment. When interaction results in scale effects it causes small problems in a breeding program because the data may be transformed and the ranking of animals will not be changed. It has been documented that genotype-environment interaction generally only accounts for a small proportion of the total phenotypic variation although strains with large genetic distance were tested under very different environmental conditions for growth rate and survival in species like salmon, rainbow trout, tilapia, rohu carp and P*enaeus vannamei.*

3

Breeding in Aquaculture

In comparison with livestock species, application of quantitative genetic principals to fish breeding has been limited until very recently. Fish farming started 4000 to 5000 years ago in China. Artificial hatching of fish was already practised in China around year 2000 BC (Lin, 1940). The earliest written descriptions of pond culture are from China during the 12th century BC (Menasveta and Fast, 1998). The development of fish culture grew parallel with the culture of silk worms, as the pupa of the silkworm and their faeces provided supplementary feed for fish (Hickling, 1962). The oldest book on aquaculture is from China by Fan Lee and was published as early as in the year 475 B.C (Ling, 1977). Fan Lee described the aquaculture techniques practised around 500 BC. It is not known when domestication of fish started in China. One can still meet carp breeders in China saying, «To get strong breeders we sample wild broodstock from the big rivers», indicating that artificial selection to improve production traits does not currently take place to a large degree. According to Schaperclaus (1961) common carp farmers in Europe have been selecting the largest fish for hundreds of years compared with thousands of years in livestock species. Some experiments to improve disease resistance and growth rate started in the 1920's (Embody and Hyford, 1925; Lewis, 1944; Donaldson and Olson, 1955).

Although fish farming is old in several parts of the world it is not obvious to what degree domestication has taken part. In Europe and Israel common carp have been bred in closed populations for a long time (Schaperclaus, 1961). Rainbow trout *(Oncorhynchus mykiss)* have been kept in captivity for many generations in Europe since the 1890's and show clear sign of tameness compared to the wild type.

Tilapia were first found in Indonesia in 1937 and farming started after the second world war in South East Asia. In Norway farming of Atlantic salmon *(Salmo salar)* started in the late 1960's and the majority of the farmed fish today has been artificial selected for 7 generations and show many signs of domestication. This effect is seen most markedly in the freshwater phase as reduced cannibalism, wildness of fry and fingerlings, which means that the behaviour of the fish has changed and they use less energy to swim, fight and try to hide behind each other.

Milkfish *(Chanos chanos)* and other species, including marine shrimp were commonly cultured on a large scale in Indonesian brackish water ponds by the 15th century AD. However, important aquaculture species like milkfish, yellowtail *(Seriola quingueradis)*, crustacean and molluscs rely largely to date (2004) on wild broodstock or fry.

Present Status of Breeding in Aquaculture

In aquaculture, selection programs are not commonly used by the industry and for many species production still rely completely on catching wild broodstock and/or fry. There is no obvious reason for the lack of efficient breeding programs in aquaculture. The economic important traits in fish and shellfish appear to be little different from those in farm animals and plants. Selection response is usually higher in fish and shellfish than in farm animals (Olesen et al., 2003).

One reason for the scarcity of breeding programs in aquaculture species is that the reproductive cycle is often complex, and so is frequently not fully understood, and is therefore not able to be completed or controlled in captivity. This is the case particularly for marine species. A further factor contributing to the scarcity of breeding programs in aquaculture species may be the deterioration of the stock simply because of a rapid build up of inbreeding as a result of using few broodstock each generation (and without identification to prevent re-use). This is a problem in all species with high fecundity. Other reasons may be that researchers, extension personnel and fish farmers are not well educated in breeding theory, since education in fish biology involves little or no attention to quantitative genetics and breeding plans. The majority of published papers discussing genetics in aquaculture deal with genetic variation in qualitative traits and population genetics.

Because there has been little interest in developing breeding programs in aquaculture, the information about phenotypic and genetic parameters of economically important traits are quite limited for most of the species farmed. Before a breeding program can be established, breeding goal must be defined, estimates of genetic variance, heritability, phenotypic and genetic correlations among traits must be available. Some of the first estimates of heritability in aquatic species were published by Aulstad et al. (1972) for body weight and body length in rainbow trout fingerlings and by Kirpichnikov (1972) for body weight of common carp fingerlings.

Later a large number of reliable estimates of genetic and phenotypic parameters were published particularly for salmonid fishes. However, data is still lacking for a number of important species. There is therefore a great need to run breeding experiments in order to get reliable estimates of genetic parameters for economically important traits in the most important farmed species.

The following chapters describe the basic principles of selective breeding programs with the aim of leading to an understanding of how knowledge of mendelian, population and quantitative genetics can be used for the permanent improvement of fish and shellfish.

Basic Genetics

Molecular Genetics

DNA – Molecule: The cell has its own metabolism, it can grow, multiply and respond to external influences. The cell is surrounded by a membrane, which makes it possible for substances to pass in and out. The nucleus of the cell is surrounded by the cytoplasm, which is a semi fluid mass and consists of organic material, particularly protein. In addition, the cytoplasm contains several particles called organelles: mitochondria, Golgi apparatus, ribosomes, lysosymes and so on. In the mitochondria there is some genetic information, but its effect is directed by the nucleus.

The nucleus is of particular interest since it contains all the genetic information about the development, metabolism and behaviour of the organism. Every cell of an individual has this complete set of genetic information stored in the DNA-molecule. The DNA is organised in threadlike bodies called chromosomes within the nucleus membrane. The chromosomes are rich in nucleoproteins, which are basic proteins that bind to nucleic acids. Nucleic acids are of two types depending of the attached sugar molecules: Deoxyribose nucleic acid (DNA) and ribose nucleic acid (RNA). For some time it was not obvious which of these was the carrier of the genetic information, until it was established that DNA is the basic substance for inheritance. In some simple viruss without DNA, RNA is the genetic material.

RNA differs from DNA in having the nucleotide base uracil (U) instead of thymine. The DNA molecule has 10 base pairs in each full turn of the helix. The distance between each pair is 3.4 Å (1 Å= ten million part of a mm). If a DNA molecule in an animal cell was stretched out, it would be more than two metres long. With different sequences of base pairs the genetic information, which can be stored is almost infinite. It has been calculated that a virus has 5000 base pairs in the DNA molecule.

The information (sequence) contained in this length of DNA could be written on one page of a book. A cell of bacteria contains 1000 times more information or 1000 pages in a book. In a mammalian cell the genetic information is so large that 1 million pages would be needed to write the information down. The DNA molecule is organised into structures called chromosomes, which appear in pairs, one chromosome inherited from the male parent and one from the female parent.

The number of chromosomes varies among species and in fact chromosome number can be used to characterise species.

A gene is a small section of DNA sequence, which usually codes for the formation of a protein. The length of sequence representing a gene may vary from a few hundred to several thousand base pairs. Although each cell of an organism contains all the chromosomes and all genes, some types of cells utilise different genes at different stages of development and in response to different stimuli.

This results in cell differention in different tissues. The estimated number of genes varies between animals, but for most vertebrates is probably it in the order of 28 000 – 34 000 (Crollius et al., 2000). In addition to the recipe for a protein, genes also consist of a promotor, which is found in front of the protein recipe and is to begin protein synthesis.

The promotor is also responsible for determining in which cells the gene should function, at what time and in what quantity. The genes are interspersed by inactive DNA segments, which are believed to make up as much as 85 - 90 % of the DNA molecule.

Replication of DNA

Before a cell divides into two identical cells, the DNA-molecule must produce a copy itself. Watson and Crick (1953) uncovered the mechanism of DNA replication. A number of enzymes are active in the replication process. One enzyme begins the process by breaking the bond keeping the base pairs of the double helix together. The strands are divided in the middle and the two helixes are separated as illustrated. At the same time as the division process is occurring, the nucleotides of the helix are replicated. These nucleotides will form a complementary helix so that each daughter DNA-molecule will consist of one intact strand from the parental double-stranded helix and one newly synthesised complementary strand. Hence, DNA replication is a semi conservative process.

There two mutation processes, which can result in non-perfect replication. These processes are base-pair substitutions and frame shift mutations. Less than 20 % of spontaneous mutations are believed to be due to base-pair substitutions. Most of the rest are frame shift mutations in which one or several nucleotides are inserted or deleted. The most serious effect of error in copying of DNA molecule is when it occurs in production of egg and sperm because it may be transmitted to the next generation.

Protein Synthesis

Proteins play a central role in all living organisms. Variation in life and life processes is in general a reflection of variation and complexity in proteins. Every protein has a special function, which depends on the protein structure. The conformation of a protein is determined by the combination of the 20 possible amino acids, which it contains. The DNA sequence is not translated directly into amino acids and subsequent protein. In fact the DNA molecule is translated into a RNA molecule before the protein synthesis begins. This RNA molecule then passes the nucleus membrane and reaches the area in cytoplasm where the protein synthesis takes place. Thus it is this messenger RNA (mRNA), which programs the formation of the protein. The flow of information in the protein synthesis follows this path:

DNA –RNA—Protein

Enzymes, which are themselves proteins, are important catalysers in the protein synthesis. These enzymes are very efficient, specialised and direct only one or a few chemical processes.

Cell Division

Cell division is an endless repetition of growth, chromosomal replication, mitosis and cytoplasmic division. The cell cycle is traditionally divided into four phases, mitosis and three phases called interphases. In the first interphase the chromosomes decondense as each daughter cell of the previous division enter this phase. This phase is a very active period where proteins are synthesised, and the chromosomes consist of only a single DNA strand. In the second phase, which lasts for some hours, the DNA is replicated. In the third phase the cell prepares for mitosis. The proteins of the mitotic spindle, which is involved in the movements of chromosomes during mitosis, are synthesised. At this stage the chromosome consists of two DNA molecules and their related proteins joined at the centromeres are called a chromatid.

Mitosis

Mitosis is divided into four sub phases (eg. Wallace et al. 1986):

1. In prophase, the chromosomes condense, the nucleoli and nuclear membrane fade from view and the mitotic spindle apparatus forms
2. In metaphase, the chromosomes are brought to a well-defined plane in the middle of the mitotic spindle, which is attached to the chromosome at the centromere
3. In anaphase, the centromere of each chromosome separates and the two daughter chromosomes of each pair travel to the opposite ends (poles) of the spindle
4. In telophase, new nuclear membrane forms around each group of daughter chromosomes, the nucleoli appear, the chromosomes decondense, and finally the cytoplasm of the cell divides to form two daughter cells.

Meiosis

The animal cell is diploid with two sets of chromosomes and genes, originally one set from the dam (mother) and one set from the sire (father). During the process of meiosis diploid stem cells give rise to the haploid gametes (eggs and sperm), but as the gametes join together at fertilization the progeny will be diploid.

Meiosis consists of two cell divisions with two prophases, two metaphases, two anaphases and two telophases. Before meiosis begins, the chromosomes are replicated into two sister chromatides. Each pair of sister chromatides remains together throughout the first meiotic division. No further DNA replication occurs in meiosis.

In prophases I, the homologous chromosomes come to lie alongside each other, so that each member of the pair consists of two sister chromatides. In metaphase I, synaptic pairs of chromosomes line up at the equatorial plane of the cell. In anaphase I, the member of each pair move towards opposite poles so that each half of the cell contains one chromosome of each type. Thus, the nuclei that form in telophase I are genetically haploid, but contain two copies of each sister chromatid. In the following interphase period there is no DNA replication.

The second meiotic division for sperm is similar to mitosis. The sister chromatides separate at their centromeres and move to opposite poles and four haploid sperms are produced.

Meiosis in females results in only one egg. The other cells are lost as polar bodies following unequal cleavages in telophase I and II. In telophase I cytokinesis (division of cytoplasm) result in producing a large cell and a tiny functionless polar body. In fish the second meiotic division will not be completed before fertilization has occurred. Sperm and egg will fuse and a second polar body will be excluded.

Location of Genes

The location on a chromosome is called a locus (plural, loci). In a diploid animal there are homologos pairs of chromosomes, each chromosome in the pair containing a complete set of genes (apart from in the case of the sex chromosomes in some animals). The alternative forms of genes or non-coding sequence at a locus are known as the alleles of the locus. There may be a great deal of genetic variation among normal alleles. Often multiple alleles can occur at a given locus although any one animal can have only two alleles - one from each parent.

Mendelian Inheritance

The Austrian monk Johann Gregor Mendel (1822-1884) discovered the basic rules of heredity. The background was a series of well planed crossbreeding experiments and mathematical calculations of the result, which he published in 1866. Mendels two laws of inheritance are:

Inheritance is by units or particles, now called genes, which maintain their identity from generation to generation. It is also known as the law of segregation of genes. These units are present in duplicate in each individual. It is also called the law of independent assortment, which means that segregation at one locus does not influence segregation at another.

Mendels statement that the inheritance is units or particles is fundamental to genetics. He thought that all genes segregated independently. Later it was shown that because of linkage not all genes segregate independently. He discovered dominance and commented on it.

When an individual reproduce it transmits to each offspring one or the other of the genes in each pair it possesses. Thus the parents give to its progeny only a sample of half of its own genes. The laws of chance govern this sampling. Mendel's findings showed that Darwins theory of blending inheritance was wrong. Darwin believed that the «blood» or traits that are inherited from both parents, blended in the offspring, just as two colours of ink blend when they are mixed.

Genes on the same chromosome are expected to be linked in inheritance. There is, however, a process by which homologous chromosomes can break at corresponding sites and exchange parts, thus leading to recombination of the genes located on these chromosomes.

Such an exchange is called crossover and is illustrated. The result of this meiotic event is that some of the offspring receive recombinant gametes from their parents. As the length of chromosome between two genes increases, the more likely it is that there will be a recombination between them.

The frequency of crossovers is therefore a function of the distance between the genes. In practise, the number of crossover events in a population of chromosomes, between two genes, can be used to estimate the distance between the genes. Double crossovers can also occur, though such events are rare between closely linked genes.

Mendelian Traits in Fish

Inheritance of Colour

There are examples of traits in fish, which follow simple Mendelian inheritance. The occurrence of albinism is a good example. Albinism has been observed in a number of species: In channel catfish by

Bondari (1984); in rainbow trout by Bridges and von Limbach (1972); in medaka by Yamamoto (1969a). These investigations have found that albinism is caused by a single recessive gene *(a)* and that presence of the dominant gene (*A*) causes a coloured body. A true albino (*aa*) rainbow trout has yellow body colour and the eyes are without melanin (appearing pink rather than the normal black). It should be mentioned that the flesh from the albino fish can be coloured reddish by feeding the fish with astaxanthine.

Inheritance of Scale Pattern

In common carp there is large variation in the scale pattern, which is controlled by genes in two loci. The S locus gene controls the scaliness and the N locus gene modifies the pattern to show a linear scale pattern. Thus there is an epistatic interaction between these two loci. There are basically four scale pattern phenotypes in common carp:

1. Scaly carp also called wild carp
2. Mirror carp with some scales scattered around the body
3. Linear carp with scales in linear arrays along the body
4. Leather carp with very few scales.

Gene Action

Each active gene has a certain effect. Today much research is directed towards studying single gene effects, and localising or mapping the genes to the chromosomes.

This type of work will contribute to our knowledge of how genes function. We do know that the function of genes is strictly regulated and they are switched up and down in different cells and at different times. Development of a zygote into an embryo and finally to a mature individual is a precisely organised process.

The cells differentiate into highly specialised tissues and organs, which maintain their diverse functions throughout the life of the individual. Thus the need for specific genes to be active varies during the life cycle.

The simplest form of gene effect is that the value of each gene is additive and that the sum effect of two genes in a locus is the average of the two. For quantitative traits of economic importance there are in general many loci behind each trait and the majority of the genes have small effects.

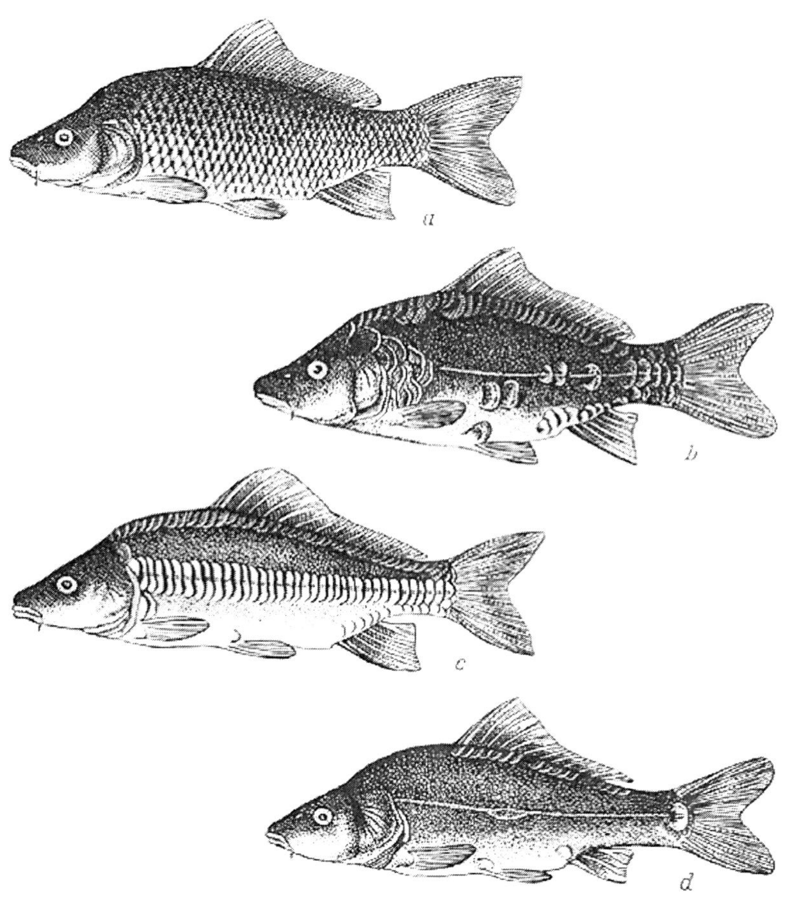

Figure: *Types of Scaling in the Common Carp. a Scaled* (SSnn *and* Ssnn*); b Scattered or Mirror* (ssnn)*; c Linear* (SSNn *and* SsNn)*; d Leather* (ssNn)*. Reproduced from Kirpichnikov (1981) by permission of Springer-Verlag GmbH.*

In many cases the effect of one allele of a gene is dependent on the effect of the other allele at the gene of the same locus, termed a dominance interaction. Let us take a lcok at the heterozygote *Aa*. There are at least three possibilities for value of this genotype as illustrated. When genotype *Aa* is equal to *AA* genotype, *A* is dominant. This interaction occurs frequently in qualitative traits, such as albino colour in rainbow trout. When the heterozygote is superior to the homozygotes (either *AA* and *aa*) the interaction is called over-dominance. When the heterozygote is average of the homozygotes there is no interaction and the inheritance is termed intermediary or additive.

Interaction between the alleles of genes at different loci is termed epistasis. This is due to that one gene in one locus influences the effect of a gene in another locus. In poultry, gene I has epistasitic control over the colour gene *C*. Birds with the following genotypes are white: *IC, Ic* and *ic* while *iC* are coloured. Birds with the dominant gene *I* is white even if they have the dominant gene *C* and birds may be white whether they are recessive in *cc* or dominant in *I* (Crawford, 1990).

The large number of genes in vertebrates, and the possibility of polymorphism at these genes, makes it evident that an enormous number of different kinds of individuals are possible. From Lush (1949) we quote « A comparison show the enormous number of different kinds of individuals possible is furnished by the physicists' estimate that the number of electrons in the universe is about 1080.

In a species in which only 200 pairs of genes are heterozygous there could be 1095 different kinds of individuals. This is a million billion times as many as there are electrons in the universe». In this example only two alleles were considered, however, if m alleles are present in each of n allelic series the number of different kinds of gametes possible is mn and the number of different kinds of genotypes.

Population Genetics

The aim of fish breeding is not to change individual fish, but rather the fish population. Thus, our knowledge about Mendelian inheritance must be extended from the level of the individual to the level of the population. Population genetics may be defined as the study of Mendelian genetics in a group of interbreeding individuals who share a common gene pool.

The number of genes in this pool depends on the species, ranging from about 4000 in the bacterium *E. coli* to about 30.000 in the vertebrate genome. Due to the accumulation of mutations over very long periods of time, the DNA sequence of these genes can vary slightly among individuals in the population.

These different forms of the genes are called alleles. If the different alleles result in differences of the gene products, the result can be genetic variation for that trait within the population. The proportions of the different alleles for each gene in the pool, the allelic frequencies, determines the genetic characteristics of the population. The product of expression of these alleles interacts with the environment in which the population exists. Some of these interactions give the individuals carrying a particular allelic conformation an advantage over other combinations. Natural selection results in an accumulation of these favourable alleles in the population and leads to a change in allele frequency.

Hardy-Weinberg Equilibrium

In 1908, the English mathematician G. H. Hardy and the German physician W. Weinberg independently discovered that in a large interbreeding population, allele frequencies will remain constant from one generation to the next (they assumed that mating was at random, and there was no selection pressure or other factors altering allele frequencies).

Such a population is said to be in Hardy-Weinberg equilibrium. To understand how Hardy-Weinberg equilibrium occurs, consider a diploid organism, in which the somatic cells have two copies of each gene located in a specific position, or locus, on the pair of homologue chromosomes received from mother and father.

Mendels first law states that in a sexually reproducing organism the two alleles separate, or segregate, during the production of gametes. Genotype frequencies may then be calculated from the allele frequencies because the proportion of an allele in a population equals the proportion of gametes containing this allele.

Thus, the genotype frequencies are entirely determined by the allele frequencies in the population. This is formulated in the Hardy-Weinberg law stating that the genotype frequencies are maintained constant in a population as long as the allele frequencies do not change and so long as mating occurs at random.

The example of skin colour in tilapia demonstrates the consequences of Hardy-Weinberg equilibrium. Pigmentation in tilapia is under control of a single gene, existing in only two allelic forms A and a. Let the frequency of A in the population be p and the frequency of a in the population be q, and $p+q=1$ (as there are only two alleles in the population).

The normally pigmented black tilapia is homozygous AA, the heterozygous Aa is bronze pigmented, and the gold individuals are homozygous aa. If the frequency of the allele A is set to 0.7 in a tilapia population, the probability of any sperm or egg (gamete) bearing an

A is 0.7, as a consequence of Mendel's first law. Then, the probability of bringing two *A*-bearing gametes together at fertilization is p x p = p2 = 0.49. So the frequency of homozygous *AA*, or black individuals, is 49 % in the population. Likewise, since the frequency of the alternative allele *a* is 1 - 0.7 = 0.3, the probability of a gamete bearing *a* is 0.3. Thus, the probability of joining two *a*-bearing gametes is q x q = q2 = 0.09. So 9 % of the fish in the population are homozygous *aa* (these fish are gold pigmented). The frequency of the heterozygous individuals (gold pigmented) is pq + qp = 2pq = 0.42, or 42%. Altogether, the frequencies of black, bronze and gold individuals add up to all of the individuals in the tilapia population, and p2 + 2pq + q2 = 1.

Figure: *Pigmentation in Tilapia. Homozygous* AA *is Black (a), Heterozygous* Aa *is Bronze (b) and Homozygous* aa *is Gold (c). Reproduced from Tave et al. (1989) by Permission.*

The picture becomes more complicated when alleles of multiple loci are considered. In the case of two allele pairs *A*/*a* and *B*/*b* of two loci located on the same chromosome, they will not segregate independently upon gamete formation. Hence, for any population not already in equilibrium, linkage between the loci will delay the attainment of equilibrium.

This is referred to as linkage disequilibria. Many generations of random mating may be required before equilibrium is attained, with more generations of mating required for more closely linked loci (as the chance of recombination between the two loci is lower).

Tests for Hardy Weinberg equilibrium are frequently used in population studies to test for random mating and for the presence of factors altering the allele frequencies. Based on the observed allele frequencies in an equilibrium population, the genotype frequencies will give an approximately fit to the Hardy-Weinberg formula. However, if the genotype frequencies deviate significantly from the expected Hardy-Weinberg values, it is necessary to investigate for evidence of:

1. Patterns of nonrandom mating, including inbreeding and crossbreeding or
2. A change in the allele frequencies due to selection, migration, mutation, or genetic Drift.

Inbreeding

Inbreeding is the production of offspring by related parents. The most extreme form of inbreeding is self-fertilization, which is possible in a few species used for aquaculture such as some species of scallop. Systematic patterns of mating, such as full-sib matings, half-sib matings, all result in inbreeding.

In a population of finite size, inbreeding accumulates over time even with random mating as by chance some related individuals will be mated. The inbreeding coefficient, F, is probability that two alleles at any locus in an individual are identical by descent (ie. both alleles can be traced back to a single common origin).

It is possible to calculate the rate of accumulation of inbreeding, ÀÛ'ÜF , both for systems of non-random mating and for a population of finite size with random mating. For example, the rate of accumulation of inbreeding per generation in a randomly mating population of finite size is:

$$\Delta F = 1/2N$$

where N is the effective number of breeding individuals, or more precisely the size of an idealized population. The idealized population is a population where (Falconer 1981):

- mating is restricted to members of the same line. In other words migration is excluded
- the generations are distinct and do not overlap
- the number of breeding individuals in each line is the same for all lines and in all generations. Breeding individuals are those that transmit genes to the next generation
- within each line, mating is random, including self-fertilization in a random amount
- there is no selection at any stage
- mutation is disregarded.

The effective population size of most aquaculture breeding populations tends to be small, as the high fecundity of aquaculture species means that relatively few broodstock are required to breed large populations. For example, with Nm males and Nf females used each generation, the effective population size is:

$$Ne = 4NmNf/Nm+Nf$$

So if a fish breeding scheme uses a mating ration of one male to two females, and 25 males are used each generation (50 females), the effective population size is approximately 67. The rate of accumulation of inbreeding in this breeding program, using the formula above, would be 0.75% per generation.

The effect of inbreeding is to increase the frequency of the homozygote genotypes. One direct consequence of this is that more recessive alleles are expressed in inbred populations. Most deleterious or disease conditions are recessive, and are therefore more likely to be expressed in inbred populations. So inbred populations therefore suffer from a reduction in fitness. Inbreeding can also lead to inbreeding depression.

Crossbreeding

In contrast to inbreeding, which increases the frequency of homozygotes, crossbreeding or the mating of unrelated individuals increases heterozygosity. By increasing the frequency of heterozygous genotypes, recessive alleles are less likely to be expressed, and so the fitness of the population can be increased (for one generation following crossing at least). Increased heterozygosity of crossbred individuals is often observed as hybrid vigor or heterosis, when the average value of the offspring for a particular trait exceeds the mean of the average values of the two parental lines. Hetercsis is exploited in some plant and animal breeding systems, but is nct widely used in aquaculture species.

An exception is Nile-tilapia in the Philippines. In a large scale trial, two different crosses, *O. aureus* x *O. spilurus* and *O. mossambicus* x *O. niloticus* showed 22 and 25% percent heterosis respectively for body weight. Another example is that hybrids of *O. mossambicus* and *O. niloticus* have higher tolerance to salinity than pure *O. niloticus* (Tayamen et al., 2002).

In contrast to genotype frequencies, allele frequencies in the population are not affected by inbreeding or crossbreeding *per se*. Other forces, such as natural or artificial selection, mutations, genetic drift, and migration, which together are responsible for evolution, do alter frequencies of alleles in the population.

Selection

Selection occurs when particular genotypes have greater reproductive success than other genotypes. For example, if a particular allele at a gene confers a reproductive advantage over other animals in the population, then the frequency of the allele in the population could be expected to increase over time.

The effect of selection can be demonstrated considering a large random mating population with a single locus, with two alleles *A* and *a* in the population (from Gardner and Snustad, 1981). Let 1-q be the frequency of *A*, and q the frequency of *a*. *A* is completely dominant in this case. A constant selective advantage is assumed for the dominant phenotypes (*AA* and *Aa*) over the *aa* genotype. Reproductive rates for genotypes *AA*, *Aa* and *aa* are 1:1:1-s respectively. The selection coefficient s, measures the disadvantage of an organism, relative to a particular reference genotype under selection, and consequently represents the intensity of selection. If *aa* is lethal (for example, Albinism in the channel catfish) then s = 1. If q in the initial population and s are both known, the frequency of *a* in each successive generation can be calculated as ;

$$\Delta q_{s} = sq^2(1-q)/1-sq^2$$

This formula can be used to show the many generations are required to reduce the frequency of lethal recessive alleles (in the case where heterozygotes *Aa* are not affected). In the example of the albinism allele above, if the initial allele frequency of *a* is 50%, it will take approximately 50 generations to reduce the frequency of the *a* allele to 5 % in the population.

Selection against the *a* allele is rapid at first, but decreases as the frequency of *a* declines. In fact, the *a* allele may never be entirely eliminated from the population through selection, as when q is small enough to allow homozygotes to occur very infrequently, the recessive allele is still harboured in *Aa* heterozygotes. In the case where heterozygotes are more viable than both homozygotes, neither allele will eliminate the other since the superior heterozygotes contain both alleles. This phenomenon is referred to as balanced polymorphism.

In nature, the selection coefficients at any one locus are generally very small, and so natural selection needs many generations to alter allele frequencies within a population. This allows genetic variation in the form of different alleles to remain within the population. As a consequence, the population may be able to respond to changes in the natural environment, if successive changes favour different alleles.

The consequence of artificially selectively breeding animals is also a change in allele frequencies of the genes which affect the selected traits are changed over time. If the trait is determined by one or a small number of genes, for example skin colour in tilapia, then the change in allele frequency due to intense selection can be rapid. This can result in rapid changes in the proportions of the genotypes.

Mutation

Mutations change one allele to another, for example *A* to *a*. Neglecting backward mutation (ie *a* to *A*), the change in frequency of *a* is proportional to the mutation rate (u) and frequency (p) of *A*. The number of generations (t) necessary for a given change in allele frequency can be computed from the equation (Falconer, 1981): pt/po = (1-u)t

where pt and po is the allele frequency in generation t and the base generation, respectively. With small per locus mutation rates, a very large number of generations are required for mutation to have an appreciable effect on allele frequencies.

A more important consequence of mutation is the creation of new allelic variants, which can be acted upon by selection. The accumulation of changes in DNA over millions of years has provided the source of variation that makes one species distinct from another, and also underlies the differences between individuals of the same species. Whereas some mutations occasionally can benefit the organism, the majority of mutations is detrimental, and are generally kept at low frequency in the population through selection.

It is possible to predict the trend in allele frequencies in the population considering both mutation and selection (s = selection intensity) simultaneously. If the frequency of the rare (and deleterious) allele is small, the change in allele frequency each generation is approximately - sq2 , while mutation increases the frequency of the rare allele (*a*) by u each generation. At equilibrium, both forces will be equal, and the frequency, which the rare allele will attain, q can be estimated as (Falconer, 1981): $-sq^2 = u$

So consider again our example of Albinism in catfish, where the *aa* genotype is lethal (s=1). If the mutation rate from the wild type to the albino allele is 10-6 each generation, and s=1, the estimated frequency which the allele will obtain in the population is 10-3.

Genetic Drift

Genetic drift is the random fluctuation in allele frequencies, which results from the sampling of alleles during breeding in small populations. The smaller the population, the greater this sampling effect will be. An extreme example is two parents both with heterozygous genotypes. If they only produce two offspring, there is a 25% chance both will be the *AA* genotype, and the *a* allele will be lost from the population. The process of genetic drift is important where the number of breeding individuals is in the tens or hundreds. Falconer (1981) noted that the direction of genetic drift could not be predicted, but it's magnitude, expressed as the variance in change of allele frequency as a result of sampling in one generation.

The effect of genetic drift is to reduce genetic variability. This reduced variability may lower the capacity of a population to adapt to changes in the environment. Pronounced genetic drift is especially likely when a small group of individuals separate from a larger population. By chance these pioneers may carry only a fraction of the genetic variability of the parental population. Thus, the gene pools of the two populations are different at the very beginning. This phenomenon is called the founder effect. Genetic drift may also occur when many members of a population die, and the few remaining individuals mate to restore the numbers of the population. This process is described as a population bottleneck. Following a bottleneck, the gene pool of the new population will be restricted compared to the original pool.

Migration

Populations may become isolated from other populations of the same species by a number of mechanisms, including physical barriers such as rivers and mountain ranges.

The processes of genetic drift and selection in different environments can cause differences in allele frequencies between the populations. If there is then some movement of individuals between these otherwise isolated populations, and these immigrants interbreed with the native population, the allele frequencies in the receiving population may be altered.

The frequency changes are proportional to the difference in allele frequencies between the populations and the rate of migration. Let us consider that a population consists of a portion of n new immigrants in each generation, and 1 – n individuals being natives. Let the frequency of a certain allele at a particular gene be qn among the immigrants and qo among the natives.

Often, immigrants will come from neighbouring populations, which may not have had time to develop markedly different allele frequencies from the population being considered. However, the import of animals from a remote population, or the introduction of sperm and eggs in domesticated population may dramatically alter allele frequencies, or indeed contribute entirely new alleles. This could be the case if some of the introduced alleles are adaptively superior in the new environment. The extreme case of such exchange occurs during interspecies hybridization (interbreeding between populations that are different enough to be ranked as separate species). Although much less common than gene flow within a species, the effects are more pronounced, as the allele frequencies may have diverged for a much longer time period. Indeed, many alleles in one species may not be present in the other (Gardner and Snustad, 1981).

Genetic Pollution

"Genetic pollution" and collateral damage from GE field crops already have begun to wreak environmental havoc. Wind, rain, birds, bees, and insect pollinators have begun carrying genetically-altered pollen into adjoining fields, polluting the DNA of crops of organic and non-GE farmers. An organic farm in Texas has been contaminated with genetic drift from GE crops grown on a nearby farm. EU regulators are considering setting an "allowable limit" for genetic contamination of non-GE foods, because they don't believe genetic pollution can be controlled.

Because they are alive, gene-altered crops are inherently more unpredictable than chemical pollutants-they can reproduce, migrate, and mutate. Once released, it is virtually impossible to recall GE organisms back to the laboratory or the field.

Damage to Beneficial Insects and Soil Fertility

In 1999, Cornell University researchers made a startling discovery. They found that pollen from GE Bt corn was poisonous to Monarch butterflies. The study adds to a growing body of evidence that GE crops are adversely affecting a number of beneficial insects, including ladybugs and lacewings, as well as beneficial soil microorganisms, bees, and possibly birds.

Creation of GE "Superweeds" and "Superpests"

Genetically engineering crops to be herbicide-resistant or to produce their own pesticide presents dangerous problems. Pests and weeds will inevitably emerge that are pesticide or herbicide-resistant, which means that stronger, more toxic chemicals will be needed to get rid of the pests. Herbicide resistant "superweeds" are already emerging. GE crops such as rapeseed (canola) have spread their herbicide-resistance traits to related weeds such as wild mustard plants. Lab and field tests also indicate that common plant pests such as cotton bollworms, living under constant pressure from GE crops, will soon evolve into "superpests" completely immune to Bt sprays and other environmentally sustainable biopesticides. This will present a serious danger for organic and sustainable farmers whose biological pest management practices will be unable to cope with increasing numbers of superpests and superweeds.

Genetic Distance and Population Divergence

As a result of the forces of genetic drift, selection and mutation, populations which are isolated from each other, can experience divergence in allele frequencies, particularly if population sizes are small or if selection is for different characteristics in the different populations (for example, different diseases). In fact the extent of the divergence in allele frequencies between populations can be used to infer how long the populations have been isolated and the extent of gene flow between them.

Nei's genetic distance has been widely used to quantify genetic differences between populations. Recent advances in molecular techniques and the identification of highly polymorphic repetitive DNA elements have made it relatively easy and cheap to genotype a large number of loci, improving the estimates of genetic distance.

4

Recent Advances of Cytogenetics in Fisheries

Cytogenetics is a branch of genetics that is concerned with the study of the structure and function of the cell, especially the chromosomes . Barbara McClintock and Harriet Creighton were the ones who worked on maize in early years popularizing cytogenetics. Fish make up half of the extant vertebrate species, they exhibit a huge level of biological diversity and, as food resource, they play major roles in the culture and economy of human populations. For most of the biologists, cytogenetics begins to be recognised as an essential tool to approach questions in both basic and applied ichthyology.

Cytogenetics-a Breif History

Modern cytogenetics is generally said to have begun in 1956 with the discovery that normal human cells contain 46 chromosomes by Tjio and Levan. This discovery was aided by a new technique of slide preparation utilising a *hypotonic* solution discovered by TC Hsu in 1952. With that advent of harvest procedures which allowed easy enumeration of chromosomes, discoveries were quickly made in abnormalities arising from nondysjunction events which cause cells with aneusomy.

In the late 1960's Caspersson developed banding techniques which differentially stain chromosomes. This allows chromosomes of otherwise equal size to be differentiated as well as to elucidate the breakpoints and constituent chromoso-mes involved in chromosome translocations. Deletions within one chromosome could also now be more specifically named and understood. Diagrams indentifying the chromosomes based on the banding patterns are known as cytogenetic maps. These maps became the basis for both prenatal and oncological fields to quickly move cytogenetics into the clinical lab where karyotyping allowed scientists to look for chromosomal alterations.. Techniques were expanded to allow for culture of free amniocytes recovered from amniotic fluid, and elongation techniques for all culture types that allow for higher resolution banding

Cytogenetics: Recent Advances, the Beginning of Molecular Cytogenetics

Fish

In the 1980s advances were made in molecular cytogenetics. FISH (fluorescence *in situ* hybridization) is a cytogenetic technique used to detect and localize the presence or absence of specific DNA sequences on chromosomes. FISH uses fluorescent probes that bind to only those parts of the chromosome with which they show a high degree of sequence similarity. Fluorescence microscopy can be used to find out where the fluorescent probe bound to the chromosomes. FISH is often used for finding specific features in DNA for use in genetic counseling, medicine, and species identification. FISH can also be used to detect and localize specific mRNAs within tissue samples. In this context, it can help define the spatial-temporal patterns of gene expression within cells and tissues.

This change significantly increased the usage of probing techniques as fluorescently labelled probes are safer and can be used almost indefinitely.

Chromosome Micro Dissection

Chromosome micro dissection is a technique that physically removes a large section of DNA from a complete chromosome. The smallest portion of DNA that can be isolated using this method comprises 10 million base pairs - hundreds or thousands of individual genes. To prepare cells for chromosome micro dissection, a scientist first treats them with a chemical that forces them into metaphase: a phase of the cell's life-cycle where the chromosomes are tightly coiled and highly visible. Next, the cells are dropped onto a microscope slide so that the nucleus, which holds all of the genetic material together, breaks apart and releases the chromosomes onto the slide. Then, under a microscope, the scientist locates the specific band of interest, and, using a very fine needle, tears that band away from the rest of the chromosome. The researcher next produces multiple copies of the isolated DNA using a procedure called PCR (polymerase chain reaction). The scientist uses these copies to study the DNA from the unusual region of the chromosome in question.

Comparative Genomic Hybridization

Comparative genomic hybridization (CGH) or Chromosomal Microarray Analysis (CMA) is a molecular-cytogenetic method for the analysis of copy number changes (gains/losses) in the DNA content of a given subject's DNA and often in tumor cells. CGH will detect only unbalanced chromosomal changes. Structural chromosome aberrations such as balanced reciprocal translocations or inversions can not be detected, as they do not change the copy number.

Virtual Karyotype

Recently, platforms for generating high-resolution karyotypes *in silico* from disrupted DNA have emerged, such as array comparative genomic hybridization (arrayCGH) and SNP arrays. Conceptually, the arrays are composed of hundreds to millions of probes which are complementary to a region of interest in the genome. The disrupted DNA from the test sample is fragmented, labelled, and hybridized to the array. Knowing the address of each probe on the array and the address of each probe in the genome, the software lines up the probes in chromosomal order and reconstructs the genome *in silico*. Virtual karyotypes have dramatically higher resolution than conventional cytogenetics. The actual resolution will depend on the density of probes on the array. Currently, the Affymetrix SNP6.0 is the highest density commercially available array for virtual karyotyping applications. It contains 1.8 million polymorphic and non-polymorphic markers for a practical resolution of 10-20kbýÿabout the size of a gene. This is approximately 1000-fold greater resolution than karyotypes obtained from conventional cytogenetics.

Fish Cytogenetics

Only limited knowledge was available in fisheries about assessing biodiversity in a river basin, to control the reproduction in a cultured species, from studying the ancestry of a given lineage to tracking down DNA sequences in a fish genome, etc. These new molecular techniques when introduced into fisheries created a revolution as there were a lot of things still to be learnt in this field. Summarized below are some of the recent findings in fish cytogenetics.

Advances

1. A comprehensive compendium of chaotic killifish karyotypes. Martin Vlker , Petr Rb , Harald Kullmann karyotyped killifish. Fish karyotypes are generally quite well conserved in evolution, with more than 50% of species having 2n=48 or 2n=50 chromosomes. However, some groups of killifishes (Cyprinodontiformes) show an enormous karyotypic variability which is most distinct in the African family Nothobranchiidae. Using a combination of cytogenetic and molecular phylogenetic methods, our studies of the nothobranchiid genus Chromaphyosemion revealed chromosome numbers ranging from 2n=20 to 2n=40, several phylogenetically independent reductions of 2n, a high diversity of NOR phenotypes, variability in the number of chromosome arms due to inversions and heterochromatin additions and possibly also independent evolutions of XX/XY sex chromosome systems.
2. Cytogenetic studies of Atlantic salmon, Salmo salar L., in Scotland. Chromosome numbers for Atlantic salmon, Salmo salar L., from ten populations in Scotland were ascertained. The majority of fish had 2n = 58, NF = 74 karyotypes, and no polymorphisms between populations were found. The findings suggest that Atlantic salmon in Scotland are cytogenetically homogeneous.
3. FISH and DAPI staining of the synaptonemal complex of the Nile Tilapia. K. Ocalewicz , 2, J. C. Mota-Velasco, R. Campos-Ramosand , D. J.Penman of Institute of Aquaculture, University of Stirling used FISH to stain synaptonemal complex of Nile fish. Bivalent 1 of the synaptonemal complex (SC) in XY male Oreochromis niloticus shows an unpaired terminal region in early pachytene. This appears to be related to recombination suppression around a sex determination locus. To allow more detailed analysis of this, and unpaired regions in the karyotype of other Oreochromis species, they developed techniques for FISH on SC preparations, combined with DAPI staining. DAPI staining identified presumptive centromeres in SC bivalents, which appeared to correspond to the positions observed in the mitotic karyotype.
4. Cytogenetic characterisation of the dwarf oyster (Ostreola stentina). The chromosomes of O. stentina were studied using conventional Giemsa staining, chromosome measurements, C-

and restriction endonuclease banding, and fluorescent in situ hybridization (FISH). Comparative analysis of the different karyotype patterns obtained for this species to those of other flat oyster for which data has been previously published was performed and provided new insights into oyster evolution and systematics within this family.

5. gene mapping of 28S rDNA sites in allotriploid Cobitis females. Among many natural Cobitis populations some polyploid hybrid forms were detected. There is a little information about genome rearrangement regarding diplo- and polyploid individuals.). Chromosomes were examined by Fluorescence in situ hybridization method (FISH) with 28S rDNA as a probe. The rDNA sites were identified in the terminal telomeric position of three submetacentric and four subtelo-acrocentric chromosomes. Moreover some of observed signals were stronger then others. One small sm/st chromosome possessing rDNA sites on both p and q arms seemed to be a marker of the karyotype of allotriploid females. This study brings useful information about polymorphism of 28S rDNA sites regarding their number, location and size in triploid specimens.
6. Mugilidae: towards a cytogenomic approach. Approx-imately 25% among the over 70 species of Mugilidae (Teleostei) have been cytogenetically analysed. Most of them show the conservative 48 uni-armed karyotype, with small differences concerning the absence or the presence of short arms on a single subtelocentric chromosome pair. The Mugil curema species complex constitutes an exception, with karyotypes mainly or exclusively composed of bi-armed chromosomes and a conserved NF=48. Moreover, preliminary data from some species with the conservative karyotype have suggested the existence of different types of satellite DNA.
7. Chromosome studies of European leuciscine fishes. Leuciscine cyprinids possess 2n = 50 and extremely uniform karyotype with 8 pairs of m, 13 -15 pairs of sm and 2-4 pairs of st/a chromosomes. The largest pair is characteristically st/a element - "leuciscine" cytotax-onomic marker. However, the interspecific homology of this chromosome pair could not be assessed due to inability to produce serial banding patterns in fish chromosomes. In a study, laser was used to dissect (10 - 15 copies of marker chromosome) whole chromosome probe (WCP)

from karyotype of roach, Rutilus rutilus to ascertain the interspecific homology of marker chromosome using cross-species in situ hybridization. WCP was hybridized to chromosomes of widely distributed species, consistently hybridized to the various proportions of distal part of longer arm indicating either sequence homology here and/or problem with production of WCP.

Non Viral Gene Delivery Approaches in Fish

Gene delivery using nonviral approaches has been extensively studied as a basic tool for intracellular gene transfer and gene therapy. Since its development, gene therapy has held great promise in the field of medicine. The ability to introduce functional genes into mammalian cells containing otherwise defective copies offers many new possibilities for the treatment of commonly acquired and inherited fish diseases such as autoimmune disorders and monogenic diseases.

The development of efficient gene therapy depends on an efficient transfer of therapeutic genes into a cell to replace or silence defective ones associated with fish disease. Viral vectors like adenoviruses and retroviruses are commonly used in gene therapy due to their high efficiency of gene delivery.

However, there are several recurring issues that have led to a reconsideration of the use of viral vectors in clinical trials, such as immunological problems, insertional mutagenesis and limitations in the size of the carried therapeutic genes.

The ultimate goal of gene therapy is to allow for continuous expression of transgenes in specific target tissues with minimal toxicity. Gene therapy has many benefits over other therapeutic methods; most importantly, it allows for selective treatment of the genetic defects which cause a disease in the specific cells or tissues that are effected. Recently, non-viral particles have been receiving increasing attention in gene therapy, since they can overcome major viral delivery toxicity issues, it remains a great challenge to find a carrier that will

(1) load genetic materials,
(2) pass the material through cellular barriers without causing a foreign body immune response,
(3) release it into the cell nucleus,
(4) Allow the visualization of this entire process without degrading the genetic materials.

Gene delivery refers to the transmission of nucleic acid (DNA and RNA) encoding a therapeutic gene of interest into the targeted cells or organs with consequent expression of the transgene. Due to electronegativity of naked DNA, the negatively charged cell membrane inhibits the entry of DNA, also the unprotected DNA will be rapidly degraded by nucleases present in plasma.

So there is a need to develop a safe and efficient gene transfection therapy system. Recently many techniques have been developed for the introduction of nucleic acids includes both viral and non-viral vectors, but viral vectors may induce a range of fish infections, so non viral gene delivery systems are getting more attention from the viewpoint of increased transfection efficiency and reduced immunogenic risks.

Methods of nonviral gene delivery have also been explored using physical (carrier-free gene delivery) and chemical approaches (synthetic vector-based gene delivery). Physical approaches, including needle injection, electroporation, gene gun, ultrasound, and hydrodynamic delivery employ a physical force that permeates the cell membrane and facilitates intracellular gene transfer. The chemical approaches use synthetic or naturally occurring compounds as carriers to deliver the transgene into cells.

The earliest chemical methods for DNA delivery were introduced in the late 1950s. These techniques used high salt concentration and polycationic proteins to enhance nucleic acid entry into the cell. Over a 10-year period starting in 1965, a variety of other chemicals were introduced, including DEAE dextran and calcium phosphate, which interact with DNA to form DEAE-dextran—DNA and calcium phosphate—DNA complexes, respectively. After the complexes are deposited onto cells, they are internalized by endocytosis

Microinjection:

Microinjection simply involves using remote-control levers to operate a micropipette tipped with a very fine glass needle. Individual fertilised eggs are held steady by suction against a blunt ended tube and the micropipette needle is inserted into the egg . A small volume of buffer (usually 1—2 nl) containing a high copy number (106—107 copies) of the transgene construct is then injected into the egg, the needle is removed and the egg incubated in the normal way. Fish eggs are surrounded by a chorion that rapidly hardens after fertilisation, making it difficult to penetrate.

Partly owing to the hardening of the chorion, it is also extremely difficult to identify the position of the chromosomes in fertilised eggs. The transgene should be injected at the single cell stage to ensure integration in all the cells of the GMO. However, even when this is done, integration (if it happens at all) often only takes place after some cleavage divisions have occurred. This produces mosaic embryos where the transgene is present in some cells but not others. One drawback of the approach, however, is that microinjection can be achieved only one cell at a time, which limits its use to applications in which individual cell manipulation is desired and possible, such as producing transgenic organisms.

Though relatively efficient, the method is also rather slow and labourious and therefore neither appropriate for research with large numbers of cells nor practical for DNA delivery in vivo. Injection can cause damage that affects embryonic survival and can result in quite high mortalities. Often fewer than 50% of eggs which have been microinjected express the transgenic DNA in early embryos and expression can decrease significantly after a few days. Better results can be obtained experimentally with smaller fish species such as zebrafish (Danio rerio) and medaka (Oryzias latipes).

Electroporation:

Electroporation is a method which was first developed for work with cells in tissue culture and it involves subjecting the cells to a short burst of electrical impulse. When cells are treated like this, their cell walls become temporarily much more porous and larger molecules than usual can pass through. Cells to be treated are suspended at high concentration in a solution of high copy number DNA construct and held in a 1—2 ml container with flat electrodes on each side. The transient current is passed and DNA construct molecules pass through the cell membrane. The electrical pulse can have a range of voltages and different rates of decay after the pulse and these are characteristics that are varied in order to produce optimum electroporation results. A major advantage of electroporation over microinjection is that there is no need to handle and manipulate eggs individually.

Treatment of cells with colcemid before they are electroporated increases the frequency of transfection. This is most likely due to the arrest of cells at metaphase and the associated absence of nuclear envelope or to an unusual permeability of the plasma membranes. Linearized DNA is far more efficient in transfection than circular supercoiled DNA.

Biolistics:

Particle bombardment, which is also called biolistic particle delivery, can introduce DNA into many cells simultaneously. In this procedure, DNA-coated microparticles (composed of metals such as gold or tungsten) are accelerated to high velocity to penetrate cell membranes or cell walls. Bombardment is widely employed in DNA vaccination, where limited local expression of delivered DNA (in cells of the epidermis or muscle) is adequate to achieve immune responses.

Because of the difficulty in controlling the DNA entry pathway, this procedure is applied mainly adherent cell cultures and has yet to be widely used systemically. In plants, where there is a tough cell wall, experimenters have resorted to brute force to get foreign DNA into the cells. Biolistic involves coating microscopic particles, usually of gold, with DNA construct and explosively firing these particles directly into the cell through the cell membrane. This method has been tried with fish, sea urchins and oysters and results suggest that viable transgenic embryos can be produced. However, there is currently no evidence that biolistics offers a more effective or more efficient method of producing transgenic fish than microinjection.

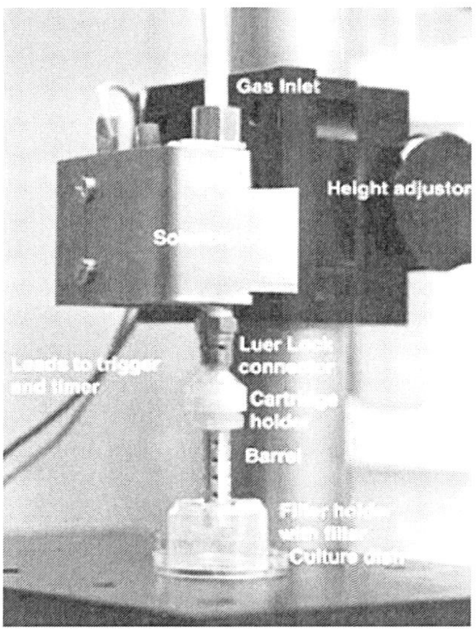

Figure:*Biolistics*

Ultrasound-Facilitated Gene Transfer

The discovery that ultrasound can facilitate gene transfer at cellular and tissue levels expands the methodology of gene transfer by physical methods. A 10- to 20-fold enhancement of reporter gene expression over that of naked DNA has been achieved. The transfection efficiency of this system is determined by several factors, including the frequency, the output strength of the ultrasound applied, the duration of ultrasound treatment, and the amount of plasmid DNA used. The efficiency can be enhanced by the use of contrast agents or conditions that make membranes more fluidic.

The contrast agents are air-filled microbubbles that rapidly expand and shrink under ultrasound irritation, generating local shock waves that transiently permeate the nearby cell membranes. Unlike electroporation, which moves DNA along the electric field, ultrasound creates membrane pores and facilitates intracellular gene transfer through passive diffusion of DNA across the membrane pores. Consequently, the size and local concentration of plasmid DNA play an important role in determining the transfection efficiency. Efforts to reduce DNA size for gene transfer by methods of standard molecular biology or through proper formulation could result in further improvement. Interestingly, significant enhancement has been reported in cell culture and in vivo when complexes of DNA and cationic lipids have been used. Since ultrasound can penetrate soft tissue and be applied to a specific area, it could become an ideal method for noninvasive gene transfer into cells of the internal organs. Evidence supporting this possibility has been presented: in one study, plasmid DNA was co-administered with a contrast agent to blood circulation, and this was followed by ultrasound treatment of a selected tissue. So far, the major problem for ultrasound-facilitated gene delivery is low gene delivery efficiency.

Calcium Phosphate Precipitation:

In this approach, the DNA preparation to be used for transfection is first dissolved in a phosphate buffer. Calcium chloride solution is then added to the DNA solution; this leads to the formation of insoluble calcium phosphate which co-precipitates with the DNA. The calcium phosphate-DNA precipitate is added to the cells to be transfected. The precipitate particles are taken in by the cells by phagocytosis . Initially, 1-2% of the cells were transfected by this approach. But the procedure has now been modified to obtain transfection of upto 20% of the cells. In a small proportion of the transfected cells, the DNA becomes integrated into the cell genome producing stable or permanent transfection.

Deae-Dextran Mediated Transfection:

DEAE-dextran (dimethylaminoethyl-dextran) is water soluble and polycationic, i.e., has a multiple positive charge. It is added to the transfection solution containing the DNA. In some unknown way, DEAE-dextran brings about DNA uptake by the cells through endocytosis. Possibly its interaction with the negatively charged DNA molecules and with the components of cell surface plays an important role.

Lipids, Liposomes and Polymers:

A number of cationic substances, such as *liposomes, lipids* and *polymers,* have been investigated for their capacity to improve efficiency of gene delivery. Cationic lipids and liposomes are the most widely used and several have been shown to raise the efficiency of *in vitro* gene delivery in many cell types. The mechanisms by which lipids and liposomes improve gene transfer is not clearly understood, but several factors, such as the type of cell and their proliferative status, appear to be involved. The positively charged lipids and liposomes are thought to improve transgene delivery mainly through binding to and condensing negatively charged DNA, forming a complex called lipoplex in which the DNA is protected against extracellular degradation.

Moreover, the positively charged lipoplex binds to the negatively charged cell surface molecules facilitating endocytosis. Once in the endosome, some lipids may destabilize the endosomal membrane and encourage the release of DNA into the cytosol, thus avoiding the lysosomal degradation pathway. Such lipoplexes can effectively deliver transgenes to myoblasts in culture and many different types of cells *in vivo*, including epithelial cells and hepatocytes. Non-ionic polymers such as poly *N*-vinyl pyrrolidone (PVP) and some co-polymers have been shown *in vivo* to enhance transduction of muscle fibres significantly. Based on the use of PVP, Rolland and colle-agues established the concept of the 'protective, interactive, non-condensing' (PINC) delivery system, and eported up to 10-fold enhancement of transgene expression over naked plasmid alone.

Polymeric Gene Carriers

The poor transfer efficiency and transient gene expression of polymer-DNA complexes continue to stimulate the development of more effective and less-toxic polymeric gene carriers. A number of polycations have been tested to transfer genes in vitro or in vivo, including poly-L-lysine (PLL) ,PEI, polyamidoamine (PAMAM) dendrimers, poly(a-(4-aminobutyl)-L-glycolic acid) (PAGA), poly((2-dimethylami-no)ethyl methacrylate) (PDMAEMA) ,chitosan etc.

Poly-L-Lysine

Poly-L-lysine has been widely used in the early days as a gene delivery carrier because of its excellent DNA condensation ability and efficient protection of DNA from nuclease digestion. However, its cytotoxicity and low transfection problems remain to be solved. Moreover, PLL-DNA complexes tend to aggregate under physiological conditions. Efforts have focused on improving the solubility of PLL by the introduction of hydrophilic groups, such as dextran and PEG. Both PEG-g-PLL and PEG-b-PLL have shown lower cytotoxicity and enhanced transfection efficiency than PLL.

Polyethylenimine

Polyethylenimine-DNA complex is another potent synthetic gene transfer vector which has the ability to transfect a wide variety of cells. The high transfection efficiency of PEI relates to its buffering capacity, which leads to the proton accumulation due to endosomal ATPase and an influx of chloride anions, triggering endosome swelling and disruption, followed by the release of DNA into cytoplasm. There are considerable differences in the cytotoxicity and transfection efficiency of branched and linear PEI. Toxicity remains one of the main concerns for PEI to be used in gene delivery. Cytotoxicity of PEI is thought to be derived from its ability to permeabilize cell membranes. Moreover, PEI is not degradable and high molecular weight PEI may accumulate in the body. Several approaches have been explored to reduce the cytotoxicity, such as PEGylation and conjugation of low molecular weight PEI with cleavable cross-links such as disulfide bonds, which can be cleaved in the reducing environment of the cytoplasm.

Chitosan:

Chitosan, a biodegradable polysaccharide composed of D-glucosamine repeating units, has been explored by several research groups as a nonviral gene carrier. Chitosan can efficiently bind DNA and protect DNA from nuclease degradation. Moreover, chitosan has good biocompatibility and toxicity profile, rendering it a safe biomedical material for clinical applications. Chitosan-DNA nano-particles can transfect several different cell types. However, the transfection efficiency is relatively poor.

Chitosan can be readily modified. For example, trimethy-lated chitosan can be prepared with different quarter-nization degree to increase the solubility of chitosan at neutral pH,or chitosan can be conjugated with deoxycholic acid to become a colloidal gene carrier.Both quaternarized chitosan and deoxylic acid-modified chitosan can efficiently transfect COS-1 cells.

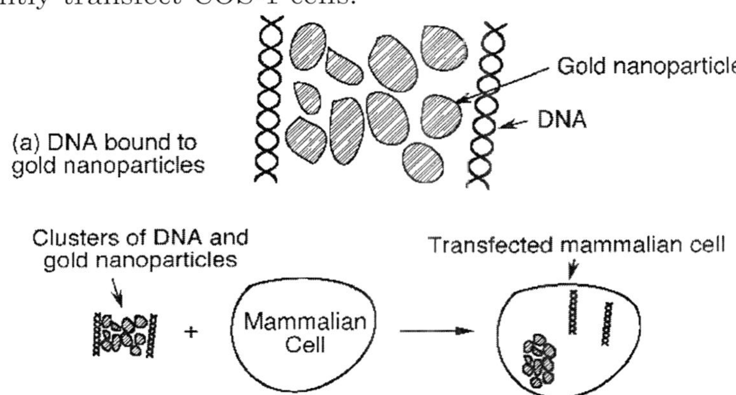

(b) Nanoparticle-mediated transfection of cells

Chitosans with different molecular weights have different DNA binding affinities and exhibit different transfection efficiencies, indicating that particle stability may be one of the rate-limiting steps in the overall transfection process. The effect of degree of deacetylation on the stability of chitosan-DNA nanoparticles is being investigated.

Polyphosphoester

Polyphosphoesters (PPE), particularly the ones with a backbone analogous to nucleic acids and teichoic acids show promise in different biomedical applications because of their biocompatibility, biodegradability, and pendent chain functionality. Several polyphosphoesters with positive charges either in the backbone or in the side chain were synthesized and evaluated as nonviral gene carriers. These polyphosp-hoesters can efficiently bind DNA and protect DNA from nuclease degradation.

Efficient transfection was found in a number of cell lines, with some of them comparable to Lipofectamine. The transfection is cell-type dependant and can be improved with the incorporation of chloroquine. All the polypho-sphoesters exhibit a significantly lower cytotoxicity than PLL or PEI in vitro and in vivo. With the structural versatility, polyphosphoester can be designed to have different hydophilicity-hydrophobicity properties. One interesting feature is that the PPE can provide extra cellular sustained release of the DNA, leading to prolonged and enhanced transgene expression in the muscle compared to naked DNA admini-stration.

DEAE-dextran and calcium phosphate methods are simple, effective, and still widely used in the laboratory for in vitro transfection. Even so, both methods are hampered by cytotoxicity and the difficulty of applying them to in vivo studies. In addition, DEAE dextran can be used neither with serum in culture medium nor for stable transfection. The calcium phosphate method also suffers from variations in calcium phosphate—DNA sizes, which causes variation among experiments. Since the early 1970s, many other polymers have been demonstrated to increase DNA uptake by cells. The most noticeable improvement is the development of artificial lipid-based DNA delivery systems. Felgner and colleagues developed the cationic lipid Lipofectin in 1987. Lipofectin—DNA complexes can be handled easily and, therefore, became one of the first chemical systems that could be used in animals and fishes.

Advantages of Non Viral Gene Delivery Systems:

- No viral components
- Low or no immunogenicity
- No limit to DNA insert size
- Cell specificity possible with targeted ligands
- Relatively simple preparation procedures
- Standardized homogenous, stable reagents
- Scale up possible

Disadvantages of Non Viral Gene Delivery Systems:

- Low transfection efficiency
- Transient gene expression-episomal expression
- Intracellular barrier- may require additional agents
- Cellular toxicity with some vectors (PEI,liposomes)
- Inflammation due to unmethylated CpG DNA sequences.

5

Algal Biotechnology in Aquaculture

Microalgae are a highly diverse group of unicellular organisms comprising the eukaryotic protists and the prokaryotic cyanobacteria or blue-green algae. The microalgae have a unique environmental status; being virtually ubiquitous in euphotic aquatic niches, they can occupy extreme habitats ranging from tropical coral reefs to the Polar Regions, and they contribute to half of the globe's photosynthetic activity. Furthermore, they form the basis of the food chain for more than 70% of the world's biomass. Microalgae are a valuable environmental and biotechnological resource, and the aim of this review is to explore the use of in vitro technologies in the conservation and sustainable exploitation of this remarkable group of organisms.

The first part of the review evaluates the importance of in vitro methods in the maintenance and conservation of microalgae and describes the central role of culture collections in applied algal research. The second part explores the application of microalgal in vitro technologies, particularly in the context of the aquaculture and biotechnology industries. Emphasis is placed upon the exploitation of economically important algal products including aquaculture feed, biomass production for the health care sector, green fertilizers, pigments, vitamins, antioxidants, and antimicrobial agents. The contribution that microalgae can make to environmental research is also appraised; for example, they have an important role as indicator organisms in environmental impact assessments. Similarly, designated culture collection strains of microalgae are used for ecotoxicity testing.

Microalgae are microscopic freshwater or marine organisms (e.g.) that play a key role in nature as a food source for higher animals (zooplankton, fish), for transferring nutrients in aquatic food webs and for balancing the exchange of CO2 between the ocean and the atmosphere. Microalgae are microscopic freshwater or marine organisms.

They are a highly diverse group, ranging in size from several hundredths of a mm to several tenths of a mm, taking many different shapes and existing singly or in chains or groups Microalgae occupy a very wide range of habitats, including forms that live in open water (phytoplankton) or on surfaces (benthic), and are adapted to extreme physical and chemical conditions (extremes of temperature, salinity, pH).

Well known natural phenomena involving microalgae include blooms of green algae in freshwater ponds or lakes during summer and "red tides" in the sea.

The diverse end user applications for microalgal biotechnology include:

- Carbon neutral Biofuel production and Bioenergy generation;
- Capture and bioconversion of carbon dioxide (CO2) from industrial processes;
- Ingredients for human health and wellbeing products, including essential omega 3 oils, antioxidants and pigments;
- Ingredients for aquaculture and agriculture feeds;
- Bioremediation and valorization of agricultural and industrial effluents.
- Energy producers and heavy industry generating CO2 as a by-product (including both fossil fuel and biomass power plants)
- Municipal waste water treatment companies;
- Intensive livestock producers releasing "point source" effluents (both agriculture and aquaculture sectors).
- Food processors releasing nutrient-enriched
- waste waters

Application of Micro Algae

Aquafeed

Microalgae are used ubiquitously as a feed source in the commercial hatchery production of juvenile marine fish and shellfish. There are thousands of marine hatcheries globally, producing billions of juvenile fish and shellfish annually. A relatively small number (~6-10) of easy-to-rear microalgae species have been adopted for this purpose. In most cases, the microalgae are cultured on site by hatchery personnel and presented live to the fish / shellfish larvae. Under this scenario, sales opportunities to hatcheries mainly consist of the equipment and consumables required for microalgae production: photo bioreactors, pumps, lights, nutrient mixes, etc. However, there is a growing trend for hatcheries to buy proprietary microalgae concentrates in order to simplify on-site operations. These concentrates are supplied by companies specialising in the large scale production and processing of microalgae.

Problems in Using Micro Algae as an Aquafeed

The high costs associated with algal production, the risks for contamination, and temporal variations in the algal food value still pose problems for any aquaculture operation depending on the mass-cultures of unicellular algae. In order to overcome or reduce the problems and limitations associated with algal cultures, various investigators have attempted to replace algae by using artificial diets either as a supplement or as the main food source. Different approaches are being applied to reduce the need for on-site algal production, including the use of preserved algae, micro-encapsulated diets, and yeast-based feeds. There is further scope to develop the sector by introducing better quality products, since it is widely acknowledged that existing concentrated microalgae products still do not match live microalgae for hatchery applications.

Nutraceuticals

The most important microalgae species for this purpose are *Dunaliella salina*, Arthrospira sp, Chlorella sp and *Aphanizomenon flosaquae*. These are mainly produced in outdoor ponds or shallow raceways, but also in closed photo bioreactors at more northerly latitudes including Europe. Certain cyanobacteria, for example *Arthrospira platensis* and *A. maxina* are also marketed as whole food, being particularly protein-rich (up to 77% dry mass) and containing all essential amino acids, a number of important essential fatty acids (EFAs) and vitamins of the B, C, D and E groups.

PUFA and HUFA

Omega 3s are natural oils of marine origin containing n-3 series long chain fatty acids such as DHA (docosahexaenoic acid) and EPA (eicosapentaenoic acid). These fatty acids are referred to as being essential in nutritional terms, since they cannot be synthesised by humans and have crucial physiological functions. Microalgae are the main source of omega 3 oils in the marine food chain, where they become accumulated especially in the tissues of oily fish such as anchovies and sardines. The major source of commercially available omega 3 oils is currently from captured marine fish such as, cod liver oil, contributing about 85% of the market by volume.

However, the supply of marinesourced omega 3 is being threatened by adverse environmental conditions that have contributed to lower DHA levels in fish oil especially from fish species from South American waters which are the major suppliers of fish oil and also depleting global fish stocks.

These adverse environmental factors coupled with depleting fish stocks is aiding the global market growth of algal based omega 3, which is currently contributing about 3% of the total omega 3 market. Microalgae based omega 3 oils furthermore appeal to vegetarian consumers and this sub-sector of the algal oils market is growing.

Effluent Remediation

It is now well known that microalgae have high potential to reduce nutrient, and organic loads from wastewaters. It is widely used for Phytoremediation in aquaculture. Removal percentages of 75%, 84% and 89% for ammonia, nitrite and phosphorous respectively have been reported.

A combination of wastewater treatment and algal carbon dioxide fixation provides incentives in the form of saving in water treatment chemicals and the subsequent environmental benefits. Furthermore, a pathway for removal of nitrogen, phosphorous and metal ions from wastewater is provided and the pathway provides algal biomass which can further be exploited for biofuel production or for other innovative products.

Probiotics in Aquaculture

Probiotics generally includes bacteria, cyanobacteria, micro algae fungi, etc. Probiotics are a cultured product or live microbial feed supplement, which beneficially affects the host by improving its intestinal balance and health of the host. The first probiotic discovered long time ago was *Lactobacillus* sp., the lactic acid producing bacteria.

Conclusion

Though there is burgeoning interest globally about biofuels from microalgae, this has been surrounded by much hype. It is becoming clearer as the sector develops that the biofuel component of microalgae will only become commercially viable if the biomass is fully exploited to utilise all of the value added components. However, at present the field of natural product processing from microalgae is underdeveloped and aside from a handful of components, there is little in the way of full cost analysis on the best products to isolate, upgrade and commercialise. This is an area which needs to receive concerted support in the short term.

Technological Constraints to Disease Prevention and Control in Aquatic Animals, with Special Rference to Pathogen Detection

Effective disease management and risk analyses rely on accurate data and information. For aquatic organisms, much of this information has been derived from a relatively narrow array of diagnostic tools, most of which are either non-pathogen-specific or have undergone "the test of time" rather than methodical validation or standardization procedures. Recently, however, this range of diagnostic methods has expanded to encompass the molecular expertise pioneered by human health and agricultural food production needs. The complexity of many of these techniques, and their rapid adaptation to "field kits" for use by non-specialist personnel, has prompted a serious re-evaluation of what we use in aquatic animal health management and why (Hiney, 1997).

This paper is aimed at determining which areas of aquatic animal health management are limited by diagnostic and pathogen detection technology, and which are adequately met by traditional methods. Specific disease examples are described elsewhere in these proceedings by the specialists working with them, thus, the points below are deliberately general and provided as "food for thought".

The Issues

Aquatic animal health management needs arise from two separate situations:

1. The proliferation of opportunistic pathogens in physiologically stressed or immunologically compro-mised host populations, requiring sensitive, early, detection of potential pathogens.

2. The spread of a primary infectious organisms between infected and uninfected individuals, stocks or populations, requiring accurate identification of the pathogens responsible for disease outbreaks, sensitive detection of pathogens in sub-clinical carriers or abnormal hosts and accurate differentiation between benign and significant infectious organisms.

Disease Diagnosis - Identification of the Cause of a Disease Outbreak.

Some diseases can be diagnosed in the field with minimal technology or the need to isolate the causative agent, e.g., bacterial gill disease in association with stressful rearing conditions (Thoesen, 1994). Others present clinical signs which defy rapid or conclusive diagnosis, e.g., Malpeque disease of American oysters, *Crassostrea virginica* (McGladdery, 1993). Yet other diseases are caused by a range of different infectious agents e.g., chitinolytic fungal and bacterial shell diseases of crustaceans (Brock and Lightner, 1990). These situations can lead to diagnostic confusion (misdiagnosis) and ineffective management. First time disease outbreaks may (and should) require sample referrals to laboratories or diagnosticians which have experience with the putative pathogen - experience often being as important as technology for rapid and accurate disease diagnosis.

What do We have?

The tools available for disease diagnosis differ between the types of aquatic organisms being examined. For finfish, there is a relatively broad range of diagnostic techniques, many of which can be used as cross checks for diagnosis of a single disease. For example, epizootic haematopoietic necrosis of redfin perch (*Perca fluviatilis*) and rainbow trout (*Oncorhynchus mykiss*) can be confirmed by: i) conventional isolation on BF-2 (bluefin gill 2) or FHM (fathead minnow) cell lines with serological identification of the iridovirus agent; ii) an indirect immunofluorescence antibody test (IFAT); iii) an enzyme-linked immunosorbent assay (ELISA); or iv) polymerase chain reaction (PCR) amplification and subsequent sequencing of the iridovirus DNA, using two published primers (OIE, 1997).

Viral encephalopathy and retinopathy (viral nervous necrosis) virus and related nodaviruses are detectable using a range of specific and less specific techniques including: i) ultrastructural confirmation of virus-induced histopathology; ii) immunohistochemistry; iii) DFAT;

iv) ELISA; and v) PCR amplification and sequence analysis (OIE, 1997). Some diseases with a more limited range of diagnostic options can be diagnosed accurately using the techniques available e.g. whirling disease of salmonids, caused by *Myxosoma cerebralis*, is presumptively diagnosed by behaviour, with confirmatory observation of the myxosporean spores in cartilage digests or histology preparations. Furthermore, few finfish diseases with a single aetiology, have defied conclusive diagnosis for long periods.

The role of multiple infectious agents in a disease can usually be resolved through experimental research and verification using Koch's postulates. One example for finfish (which has an as yet unidentified aetiologic agent) is erythrocytic inclusion body syndrome (EIBS). Although the causative agent is believed to be viral in nature, secondary infections by bacteria and fungi can confound diagnosis (Thoesen, 1994; Jarp *et al.*, 1996).

Once the primary infectious agent for such diseases is identified, subsequent diagnosis is simplified and, generally, ignores the presence of the secondary pathogens. The classic example of one such multi-factorial disease is epizootic ulcerative syndrome (EUS) which is described in detail elsewhere in these proceedings.

The range of diagnostic techniques available for molluscan and crustacean diseases is narrower than that for finfish. Most significantly, aquatic invertebrates lack self-replicating cell lines for isolation and identification of intracellular pathogens. Finfish cell lines can be used, but the nature of the isolated viruses is often subject to question, since they could be vertebrate contaminants rather than primary invertebrate pathogens (Hill *et al.*, 1986). In addition, Koch's postulates have rarely been fulfilled or replicated for molluscan or crustacean isolates from fish cellm lines.

Thus, most intracellular infections which cause overt disease in crustaceans and molluscs require histopathology for presumptive diagnosis, with ultrastructural confirmation of viral or bacterial aetiology. Histology, although labourious, has the advantage that it provides a permanent record of the pathogen *in situ* and can be used to assess focal or systemic histopathology. Conversely, it is limited in sensitivity to infectious agents which can be detected, and identified, at the light microscope level, eliminating most viruses, many bacteria, protists and even some metazoan parasites (which require whole mount or adult-stage identification).

As with finfish, multiple diagnostic techniques are available for a number of shrimp diseases (Lightner, 1996) but most clinical cases can be presumptively diagnosed using non-specific techniques (gross observation, histology and tissue smears). Confirmatory diagnosis is then achieved using culture e.g. crayfish plague (Alderman and Polglase, 1986) or electron microscopic examination of ultrastructural features (Lightner, 1996). Pathogen culture is rarely used for diagnosis of molluscan diseases with the exception of two groups of significant disease agents: i) *Perkinsus* spp. (Ray, 1966; Gauthier and Vasta, 1993; LaPeyre *et al.*, 1993); and ii) *Labyrinthuloides*-like protists (Bower, 1987; Kleinschuster *et al.*, 1998). In clinical cases, however, these are also readily diagnosed using standard histology. Most other significant disease agents of molluscs are difficult to culture, but can be isolated under certain conditions (Hervio *et al.*, 1993; 1995) i.e. acute infections. However, pathogen isolation is normally reserved for development of more specific detection and identification techniques rather than clinical diagnoses.

What are the Limitations?

Speed of diagnosis is always a concern, especially with acute losses relying on histopathology, ultrastructural confirmation or long periods of tissue/media culture. The time span required for confirmatory diagnosis is frequently overcome by remedial action being based on presumptive diagnoses such as tissue smears, gross pathology or behavioural changes. This is most effective in areas with a well defined history of the disease e.g. Denman Island disease which is caused by *Mikrocytos mackini* in Pacific oysters (*Crassostrea gigas*) on the west coast of Canada (Bower, 1988).

First time disease outbreaks in new species to culture, or appearing at a location for the first time, can undergo protracted periods of non-diagnosis or, worse, misdiagnosis. Examples include a serious disease of hard shell clams, *Mercenaria mercenaria*, caused by an unidentified *Labyrinthuloides*-like organism, "QPX". This may have been causing mortalities in pre-culture history (Drinnan and Henderson, 1963) but its significance was not fully realised until hatchery broodstock and cultured stock on grow-out beds started to die (Whyte *et al.*, 1994; Ragone Calvo *et al.*, 1997; Smolowitz *et al.*, 1996). Similarly, infectious salmon anaemia of Atlantic salmon (*Salmo salar*) was described as haemorrhagic kidney syndrome (HKS) when first detected in Atlantic Canada (Getchell, 1997). It took over a year before the causative agent was recognised as a virus and identified as ISAV, an agent previously known only from Norway (Hastein, 1997).

Where can Molecular Techniques Enhance Disease Diagnosis?

Significant pathogens that require long, complex culture or histology-based confirmatory diagnosis are prime candidates for rapid, pathogen-specific diagnostic kits. This applies predominantly to microbial pathogens, but may be equally appropriate for protists which are difficult to distinguish morphologically at the light microscope level or which have a diverse host-range. Rapid, pathogen-specific diagnostics would be particularly appropriate for disease management and control when diseases emerge in new geographic locations or host species, as described under limitations. An additional application for molecular techniques is for research into the pathogenesis of a disease via non-lethal sampling e.g. of haemolymph, fin- or gill-clips. This would provide useful information on pathogen proliferation, haemolymph profiles etc. but negates examination of the physical host-pathogen interface.

Pathogen Screening - Detecting Infectious Agents in Sub-Clinical or Healthy Organisms.

Screening for infectious agents in healthy hosts is probably the most controversial area of aquatic animal health management. This is due to: i) inconclusive negative results; ii) the potential disease risks; and iii) the difficulty of controlling a disease outbreak in naïve and/or open-water populations. Since pathogen screening is frequently a pivotal part of disease risk assessments prior to transboundary transfers, the techniques used can also be "politically sensitive".

More recently, pathogen and/or disease screening is being used to define aquatic zones (intra-national and, more rarely, international) based on health profiles. This is usually limited to specific pathogens of commercially important host species (OIE, 1977). These zones can then be used for management purposes, to allow movement of pathogen carriers between non-confluent waters where the pathogen is known to occur ("like-to-like" transfers). Pathogen detection in healthy (carrier) hosts never assures 100% confidence, statistically, therefore negative samples all have a level of error which can be directly related to the sensitivity of the screening technique(s) applied.

Since disease risk analyses have been, and continue to be, well-debated (DeVoe, 1992), more effort has been focussed on epidemiological principals in an effort to quantify and standardise the broad range of qualitative-based risk evaluations (Hiney, 1997; Thorburn, 1999).

This has revealed a broad gap between the probability of detection of a single pathogen in a given sample and the statistical confidence in that detection. This reflects non-survey-based assumptions for pathogen prevalence in wild or open-water populations, as well as detection sensitivity, since prevalence is one of the critical factors determining confidence of detection of a single pathogen in any given sample size (Ossiander and Wedemeyer, 1973).

What do we have now?

Pathogen screening of aquatic animals involves the same techniques described above for disease diagnosis. In addition to culture-based techniques, immunoassays (fluorescent antibody tests, agglutination tests, ELISA) and nucleic acid probes have been available for finfish pathogens for years, and some form the basis of kits now used for pathogen screening (e.g. *Aeromonas salmonicida* which causes furunculosis, and *Renibacterium salmoninarum* which causes bacterial kidney disease). Pathogen-sensitive techniques for molluscs pathogens have been developed more recently e.g. immunoassays for *Perkinsus marinus* (Dungan and Roberson, 1993), *Bonamia ostreae* (Mialhe *et al.*, 1988), *Vibrio tapetis* (causative agent of brown ring disease of Manila and Portuguese carpet clams, *Ruditapes philippinarum* and *R. descussatus*, respectively) (Castro *et al.*, 1995) and giant rickettisia of the sea scallop *Pecten maximus* (Le Gall *et al.,* 1992).

However, none of these techniques has yet been transformed from research to routine diagnostic application and histology remains the most broadly used detection/diagnostic method applied to molluscs.

Detection of sub-clinical infections in shrimp is limited to infectious hypodermal and hematopoietic necrosis virus (IHHNV), using bioassays as well as *in situ* and dot blot hybridisation or PCR of viral product in haemolymph or tissue homogenates, and baculoviral midgut gland necrosis virus (BMNV) using bioassays in susceptible *Penaeus japonicus* and a fluorescent antibody test (Lightner, 1996). Stress-induced enhancement of infections is another procedure used to enhance some sub-clinical viral infections in shrimp (and other aquatic species) which cannot be detected by histology (Lightner, 1996).

What are the Limitations?

For cryptic infectious agents (mainly microbial) routine diagnostic procedures on healthy animals, especially non-culture-based techniques, are particularly weak in detection sensitivity. Thoesen (1994) lists several diseases caused by primary pathogens for which there are no procedures documented for detecting sub-clinical infections (e.g. *Vibrio salmonicida,* cold-water vibriosis or "Hitra disease"; channel catfish virus, CCV; *Haplosporidium nelsoni,* MSX; and *H. costale,* SSO, of American oysters, *Crassostrea virginica*). Lightner (1996) lists very few pathogens of shrimp for which there no methods to detect sub-clinical infections (hepatopancreas parvovirus, HPV). However, some pathogens can only be detected following stress-testing (e.g. monodon baculoviruses, MBV).

For most disease agents in sub-clinical or abnormal "carrier" hosts, this means that sample size or sampling frequency has to be increased to enhance the level of confidence in detection. For techniques such as histology and ultrastructure, this frequently involves compromise between sample size (confidence level) and resource capability (time and manpower). For other more sensitive techniques (tissue culture, immunoassays and nucleic acid probes), the compromise may involve time, expense and specialist resource factors.

Where can Molecular Techniques Enhance Zonation Establishment and Surveillance or Transfer Disease Risk Analysis?

As described above, most agents of significant infectious diseases are difficult to detect using routine diagnostic techniques in healthy, sub-clinical hosts. This means that establishing an area which is designated free of a specific pathogen, inherently, includes a degree of error. Molecular screening techniques for specific pathogens could reduce this error margin by increasing confidence of detection. This would be especially important for areas that export live aquatic animals on a regular basis ("uninfected" zone to "uninfected" zone transfers). However, the pathogen specificity of these screening techniques negates detection of any other pathogenic or potentially significant organisms in the same specimens. Additional non-specific, but less sensitive, screening techniques may, therefore be required to give a true health "profile". In addition, full-scale molecular-based testing of populations for a given pathogen, especially where there has been no history of the disease, could meet with varying degrees of resistance on both a practical and political level. Interpretation of low positive results from such an area would be especially problematic and difficult to resolve.

In conclusion, molecular techniques might best serve as confirmatory screening to reinforce/refute results from general screening methods both for establishing zones and for certifying stocks free of specific pathogens. This would reduce the sample size and frequency required for high-technology screening, making their application more practical and easy to justify. Ideally the confirmatory screening should be on the same specimens (or sub-samples) from the same collections to ensure cross-reference validity. Certification of stocks as free of a specific pathogen could also benefit from the application of molecular-based detection techniques, especially where transmission is direct and negative result accuracy is imperative to prevent the spread of an endemic disease. Again, however, use of molecular probes or other pathogen-specific assays would mean any other infectious organisms would be undetected. Thus, as with zonation and surveillance, this pathogen-specific technology may best be applied as a confirmatory detection method, especially for certification of transfers from disease endemic areas.

Epizootiological Research - Determining the Factors that Trigger Pathogen Transmission and Proliferation.

Since pathogen eradication is rarely achieved in open-water or flow-through production systems, this is a crucial area of scientific research. It provides the information essential for effective reduction of disease losses to a negligible or economically acceptable level. Epizootiology is a complex science, involving detailed research into host immunity, physiology, genetics and environmental influences. Therefore, it requires a complex battery of techniques that range in application from controlled laboratory experiments to field observations.

What do We have now?

The methods available for epidemiological research are the same as those described above for pathogen screening and disease diagnosis

What are the Current Limitations to Epizootiological Research?

The difficulty of direct observation, handling stress and duplication of environmental variables in laboratory investigations often complicates the process of quantifying qualitative clinical and sub-clinical disease observations. In addition, despite extensive and well studied physiological and immunological parameters for finfish host-environment-pathogen research (Thomas and Woo, 1995; van Muiswinkel, 1995), a lack of standardisation and validation of routine diagnostic procedures has negated their direct application to epidemiological investigations of finfish diseases (Klontz, 1993; Thorburn, 1999). Research into host-pathogen interactions is further complicated for molluscs and crustaceans, where molecular immunology has only come under close scrutiny relatively recently (Bachere *et al.*,

1995). One notable exception is research on Gaffkemia (caused by *Aerococcus viridans* var. *homari*) of lobsters (*Homarus americanus*) (Stewart and Zwicker, 1972; 1974). Sadly, however, this case bears little extrapolation to other crustacean host-pathogen interactions since lobster and bacterium have a rather unique association, as summarised by Stewart (1984).

Another limitation to epidemiological research into aquatic animal pathogens is the inability to easily detect abnormal hosts (carrier, reservoir, accidental) of significant pathogens, especially those with low or unknown host-specificity. Abnormal hosts may demonstrate non-characteristic lesions or harbour the agents in tissues that are not infected in the "normal" host. This makes both detection and identification difficult, or even impossible, using routine diagnostic techniques.

Where can Molecular Techniques Enhance Epizootio-logical Research?

As for screening, detection of sub-clinical (pre- and post-clinical) infections is imperative for understanding the dynamics of the pathogen, the factors that trigger pathogenicity and determining optimum management strategies. This includes detection of the pathogen in the environment or "abnormal" host species. In order to improve confidence in screening such samples, pathogen-specific detection or isolation techniques are required. To date, few probes which show consistent sensitivity have been developed for such broad screening (Hiney, 1997) and this appears to be an area which merits further study and development (Stokes *et al.,* 1997).

Conclusion

In general the range of tools available and under development show different advantages and disadvantages for a range of different aquatic animal health applications. No one technique shows a replacement advantage over another, and none appear sufficient to merit "stand-alone" application, with the possible exception of pathogen-specific research.

DNA-Based Diagnostic and Detection Methods for Penaeid Shrimp Viral Diseases

The most important diseases, in terms of economic impact, of cultured penaeid shrimp in Asia, the Indo-Pacific, and the Americas have infectious etiologies. Among the infectious diseases of cultured shrimp, certain virus-caused diseases stand out as the most significant. Since the first report of a penaeid shrimp virus disease by Couch in 1974, at least 20 more viruses have been described from the penaeids (Table 1). The earliest diagnostic methods developed for these pathogens included the traditional methods of morphological pathology (direct light microscopy, histopathology, and electron microscopy), as well as enhancement and bioassay methods. While tissue culture is considered to be a standard tool in medical and veterinary diagnostic labs, it has never been developed as a useable, routine diagnostic tool for shrimp pathogens. Likewise, there are few antibody-based diagnostic tests available for the penaeid virus diseases (Lightner and Redman, 1998). The need for rapid and sensitive diagnostic methods has led to the application of modern biotechnology to penaeid shrimp diseases. DNA-based detection methods for the most important viral pathogens (IHHNV, HPV, SMV, TSV, YHV, GAV/LOV, WSSV, MBV, and BP) have been reported in the literature and some DNA-based diagnostic methods are commercially available. PCR or RT-PCR methods are available for several of these viruses and some are in routine use by certain sectors of the industry. For others, specific DNA probes tagged with non-radioactive labels provide highly specific detection methods for application in dot blot formats with hemolymph or tissue extracts, and with routine histological sections using *in situ* hybridization (Lightner, 1996; OIE, 1997; Lightner and Redman, 1998).

The OIE Fish Disease Commission at its September 1998 meeting voted to recommend to the OIE that three of the penaeid shrimp viruses diseases from the "listed" category be upgraded to the "notifiable" category. If the recommendations of the Fish Disease Commission are approved by the OIE's General Assembly in May 1999, the notifiable and listed viral pathogens of crustacea will be those shown in Table 2. All of these viral diseases affect cultured penaeid shrimp. Before a disease may be included on the OIE lists of notifiable and listed diseases, several criteria must be met: 1) the etiological agent must be known, 2) reliable diagnostic(s) methods must be available, and 3) the disease must be a significant disease of local, regional, or international importance.

The accompanying Tables in this report list the known penaeid shrimp viruses and summarize the available traditional and DNA-based diagnostic and detection methods available for the recently proposed OIE notifiable and listed pathogens of penaeid shrimp (OIE, 1997; OIE, unpublished report). The diagnostic and detection methods for these viruses are listed in Table 3. There is a growing need to standardize and validate the DNA-based diagnostic methods and the laboratories that use them. Standardization of DNA-based diagnostic methods is almost inherent in the nature of the tests. That is, a specific DNA probe, or a specific set of primers, that is used to demonstrate the presence of absence of a unique DNA or RNA sequence does not vary from batch to batch. Hence, with proper controls, these DNA-based methods are readily standardized (Reddington and Lightner, 1994). However, despite the growing dependence of the shrimp culture industry on DNA-based diagnostic methods, none of the tests that are available from commercial sources nor from the literature have been validated using controlled field tests. Likewise, there are no formal accreditation or certification programs yet in place to assure that test results from technicians and laboratories running the tests are indeed accurate and properly controlled (Lightner and Redman, 1998). The implementation of a formal program by appropriate international agencies or professional societies is needed to validate new diagnostic methods and to periodically review the accreditation and certification of diagnosticians and diagnostic laboratories. The establishment of regional reference laboratories for DNA-based diagnostic methods of penaeid shrimp/prawn pathogens would fit well into such a program with the goal of making these methods uniform, reliable, and readily applicable to disease control and management strategies for viral diseases of cultured penaeids.

Application of Polymerase Chain Reaction for Detection of Shrimp Pathogens in India - Indrani Karunasagar

Rapid detection of pathogens would be very essential for effective health management in aquaculture. While conventional microbiological isolation methods are used in case of bacterial pathogens, histopathology is widely used to detect viral infections. However, these methods are time consuming and lack sensitivities to detect latent pathogens. Polymerase chain reaction (PCR) on the other hand is highly sensitive and rapid and can be used to detect latent pathogens.

In our laboratory, we have been studying the application of PCR for rapid detection of shrimp pathogens such as whitespot syndrome virus (WSSV) and *Vibrio* spp. WSSV is a serious pathogen that has caused extensive mortalities in shrimp culture systems in India (Karunasagar *et al.*, 1997, 1998; Karunasagar and Karunasagar, 1999). We have studied the application of PCR primers reported by Lo *et al.*

(1996) for detection of WSSV. Our results indicate that one step PCR is able to detect infection in case of clinically symptomatic animals. Two step PCR was necessary to detect WSSV in clinically asymptomatic shrimps and in other carrier animals (Otta *et al.*, 1999). Using two step PCR, a large number of apparently healthy *P.monodon* post larval stages were screened for the presence of WSSV. Only 5% apparently healthy PL gave positive reaction in one step PCR whereas 48% showed positive reaction in two step PCR (Otta *et al.,* 1999). These results suggest that two step PCR is very essential for detection of WSSV in asymptomatic animals and carriers.

Using PCR, WSSV could be detected in carrier animals such as crabs, *Acetes* spp and even in water samples. An evaluation of the relation between PCR positivity and infectivity was studied in the case of *Penaeus monodon* showing clinical signs of white spot syndrome. Viral extracts from freshly harvested shrimp (22-25g) were highly infective causing clinical signs and mortality in healthy shrimp within 48 h. Viral extracts prepared from clinically symptomatic animals, which were stored under frozen conditions (-20°C) for two months, showed positive reaction in PCR.

However, the viral extracts from the frozen specimens failed to induce clinical signs or mortality in healthy animals. This suggests that WSSV might lose infectivity during frozen storage in shrimp tissue. Therefore, PCR positivity in frozen shrimp should be interpreted with caution with respect to its potential to spread the virus.

In our laboratory, PCR is also being used to detect *Vibrio parahaemolyticus* in shrimp. Comparision of culture and PCR methods show that PCR would be very useful in detecting this organism and the technique has potential to detect even atypical strains showing variation in biochemical reactions (Karunasagar *et al.*, 1997). Thus PCR would be a very useful tool for rapid and sensitive detection of pathogens in shrimp. The technique has applications in diagnostic laboratories, for monitoring health, to identify environmental reservoirs of infections, to detect the presence of pathogens in animals in quarantine etc.

The Need for Molecular Tools in the Study of Mollusc Pathogens

The OIE list of notifiable diseases of molluscs and the pathogens causing them comprises marteiliosis (*Marteilia refringens, Marteilia sydneyi*), bonamiosis (*Bonamia ostreae, Bonamia* sp.), mikrocytosis (*Mikrocytos mackini, Mikrocytos roughleyi*), haplosporidiosis

(*Haplosporidium nelsoni, Haplosporidium costale*) and perkinsosis (*Perkinsus marinus, Perkinsus olseni*). This paper gives a brief overview of the areas in which molecular tools are needed to overcome problems associated with these diseases, and considers the needs of the Asian region.

Infection Levels

Moderate to heavy infections with *Marteilia refringens, Marteilia sydneyi,* both *Bonamia* spp., and both *Haplosporidium* spp. are relatively easy to detect by routine histology. Both *Perkinsus* spp. can be cultured using Ray's Fluid Thioglycollate Medium (RTFM) (Bushek *et al.,* 1994), allowing light infections to be amplified, and consequently detected. *Mikrocytos* spp. are much harder to detect. *Mikrocytos roughleyi* infects the haemocytes of Sydney rock oysters (*Saccostrea commercialis*) in eastern (Georges River) and western (Carnarvon, Albany) Australia, and ultrastructurally it has a single mitochondrion.

Mikrocytos mackini infects connective tissue cells of Pacific oysters (*Crassostrea gigas*) off the coast of British Columbia, and ultrastructurally it lacks a mitochondrion. Currently it is thought that these two pathogens are not closely related, and *M. roughleyi* may be more closely related to *Bonamia* spp. Macroscopically both *Mikrocytos* spp. produce macroscopic pustular or abcess-like lesions in cases of heavy infection, and if occurring in the known range of these two pathogens, may allow presumptive diagnosis. However, microscopically both *Mikrocytos* spp. are only ~2?m in diametre, do not stain well, consequently they are difficult to detect in moderate to light infections (Hervio *et al.,* 1996). Therefore probes are needed to detect light infections with all OIE listed pathogens except (*Perkinsus* spp.), and light to moderate infections with *Mikrocytos* spp.

The Identification of Species

The inter-relationships of all the OIE listed notifiable pathogens are currently uncertain. *Marteilia sydneyi* was initially distinguished from the previously described *Marteilia refringens,* on the grounds that the latter possessed refringent granules, whereas the former did not. However, *M. sydneyi* does possess refringent granules (Roubal *et al.,* 1989). Also it is unclear how the *Marteilia* sp. that caused a massive epizootic in calico scallops off the coast of Florida (Moyer *et al.,* 1993) relates to described species. Although distinction on the basis of cleavage patterns during development seems to overcome these uncertainties, the distinction of *Marteilia* spp. is still being questioned (Bower *et al.,* 1994). The two species of *Bonamia* have also not yet been clearly distinguished, although *B. ostreae* has dense forms that are seldom seen in *Bonamia* sp, and the latter has a vacuolated stage (Hine, 1991) that has not been reported from *B. ostreae*. The two species of *Haplosporidium* can be distinguished by spore size, and *H. nelsoni* can be distinguished from all other *Haplosporidium* spp. as it sporulates in epithelial cells of the digestive diverticulae, and the other species sporulate in connective tissue. Currently, *H. costale* cannot be distinguished reliably from other *Haplosporidium* spp. except *H. nelsoni*. *Mikrocytos* spp. do not resemble each other closely and *M. roughleyi* may be more closely related to *Bonamia* spp.. *Perkinsus olseni* shows similarities to *Perkinsus atlanticus* (Hamaguchi *et al.,* 1998), but *Perkinsus marinus* also shows similarities to *P. atlanticus* (Robledo *et al.,* 1997). Specific probes are currently available, or are being developed for, *Marteilia refringens, Marteilia sydneyi, Bonamia* spp., *Haplosporidium nelsoni,* and *H. costale.*

Life Cycles

Bonamia spp., *Mikrocytos mackini,* and *Perkinsus* spp., transmit directly from one host to another. *Haplosporidium* spp. and *Marteilia* spp. cannot be transmitted directly from one to another, and probably require an intermediate host (Roubal *et al.,* 1989; Berthe *et al.,* 1998). Probes currently developed or being developed for *Haplosporidium* spp. and *Marteilia* spp. will be used to identify the DNA of these pathogens in likely alternative hosts, such as filter feeding or detrivorous invertebrates.

The Asian Region

The OIE listed diseases of molluscs mainly infect bivalves in temperate regions. This is true of *Bonamia* spp. in temperate oysters (*Ostrea, Tiostrea, Crassostrea*), *Mikrocytos* spp. in temperate oysters (*Crassostrea, Saccostrea*), and *Haplospo-ridium* spp. of temperate oysters (*Crassostrea*). *Marteilia refring-ens* is also a parasite of temperate bivalves (*Ostrea, Tiostrea, Crassostrea, Mytilus, Cerastoderma*), and although *Marteilia sydneyi* occurs in the subtropics of southern Queensland, it is primarily a parasite of temperate oysters (*Saccostrea*). *Marteilia refringens, Bonamia ostreae, Mikrocytos mackini, Haplosporidium costale* and *Perkinsus marinus* have not been reported from the Asian region.

Marteilia sydneyi, Bonamia sp., *Mikrocytos roughleyi* and *Perkinsus olseni* have been reported from Australia, and a *Perkinsus* sp. from clams (*Tapes philippinarum*) in Japan (Hamaguchi *et al.*, 1998). The Japanese isolate had sequences intermediate between *P. olseni* in Australia, and *P. atlanticus* in Manila clams (*Tapes philippinarum*) from Spain. *Perkinsus atlanticus* from *Ruditapes philippinarum* and *Ruditapes decussatus* around Spain and Portugal, are closely related to *P. olseni* (Robledo *et al.*, 1997). It may be therefore that *P. olseni* was moved from Asia in Manila clams to Europe, where it was described as *P. atlanticus*. If so, *P. olseni/ atlanticus* may be widely distributed throughout Southeast Asia.

Although *Haplosporidium nelsoni* has not been formally reported from Asia, a *Haplosporidium* sp., similar in size and pathology to *H. nelsoni*, occurs in Pacific oysters (*Crassostrea gigas*) in California and in Matsushima Bay, Japan, from which the Californian stocks derived (Friedman *et al.*, 1991; Friedman, 1996). A sensitive and specific probe for *H. nelsoni* (Stokes and Burreson, 1995) labels the Californian and Japanese *Haplosporidium,* suggesting that it is also *H. nelsoni*, and that *H. nelsoni* was originally introduced into California in Japanese Pacific oysters. Therefore, some of the temperate OIE listed diseases occur in Australia and Japan, and it is likely that *Haplosporidium nelsoni* and *Perkinsus olseni/atlanticus* occur more widely in Asia than is currently realised. As bivalve health expertise becomes more widespread in Asia, other serious diseases of tropical, as well as temperate bivalves, are likely to emerge. One such pathogen may be a *Haplosporidium* sp. pathogenic in silverlip pearl oysters (*Pinctada maxima*) in northwestern Australia (Hine and Thorne, 1998). Also, an apparently infectious disease that has caused massive mortalities among akoya pearl oysters (*Pinctada fucata*) in Japan since 1994 (Miyazaki *et al.*, 1998) may well prove to be a serious disease in Asia. Currently a parasite related to *Marteilia,* called *Marteilioides chungmuensis*, which parasitizes the ova of Pacific oysters (*Crassostrea gigas*) is having a serious impact on oyster production, and may also prove to be a problem in Asia.

Once such problems have been identified, molecular tools will need to be developed to detect low infection levels, distinguish species and study life cycles, as for the currently listed diseases.

The OIE protocols are designed to control spread of aquatic animal diseases, using a certification and reporting system that requires a national infrastructure, based in law, and a network of skilled and experienced aquatic animal health specialists, including technicians, inspectors and pathologists.

Development of such systems occurs as industries to be serviced develop, and therefore are found where aquaculture industries are well established. Such established industries use hatchery production or natural spat settlement as their source of stock.

Certification allows stock to be traded between farms, and the identification and establishment of disease free zones minimizes risk of disease spread.

In developing bivalve farms, it is often necessary to initially acquire brood stock from the wild, or to move bivalves into an area to enhance natural spat settlement. Such movements are already taking place throughout Asia. This process must be undertaken with extreme care to minimize the risk of introducing disease onto the farm site. Studies on the parasites and diseases of wild bivalves in northwestern Australia have shown that the prevalence of potentially serious diseases in wild stocks may be extremely low (Table 1). Prevalences of ~0.1% are common. To detect such infection with 95% confidence, it is necessary to sample 2,994 animals, and even then a light infection may well be missed. Molecular tools are needed to detect such low levels of infection before stocks are moved.

DNA-Based Studies on Aphanomyces Invadans, The Fungal Pathogen of Epizootic Ulcerative Syndrome (Eus)

Background

Epizootic ulcerative syndrome was defined at the DFID Regional Seminar on EUS in Bangkok in 1994 as "a seasonal epizootic condition of freshwater and estuarine warm water fish of complex infectious aetiology characterised by the presence of invasive *Aphanomyces* infection and necrotising ulcerative lesions typically leading to a granulomatous response" (Roberts *et al.*, 1994). Research requirements identified from the meeting included work to compare and speciate fungal isolates from EUS outbreaks, develop diagnostic tests, and study the factors that affect transmission of the disease. Attempts are being made to apply DNA-based techniques to each of these study recommendations, and these are discussed here.

Molecular Characterisation of EUS-Associated Aphanomyces *Isolates*

Traditionally, Oomycete fungal isolates are speciated primarily on the basis of the morphology of sexual structures. However many strains, and the more pathogenic strains in particular (including

Aphanomyces astaci and *Saprolegnia parasitica*), are reluctant to produce sexual structures in culture. Sexual structures have not been demonstrated for the EUS *Aphanomyces* pathogen, and therefore alternative methods of characterisation have been used (Hatai and Egusa, 1978; Willoughby *et al.*, 1995; Callinan *et al.*, 1995; Lilley, 1997). Recently, DNA-based methods have also been applied.

Molecular data sets are rapidly becoming an essential part of any detailed fungal taxonomic study. Previous studies on other *Aphanomyces* and *Saprolegnia* species have used restriction fragment length polymorphism (RFLP) analyses to demonstrate inter-specific relationships (Yeh, 1989; Molina *et al.*, 1995) and random amplification of polymorphic DNA (RAPD) studies have been used to show detailed intra-specific lineages (Huang *et al.*, 1994; Malvick *et al.*, 1998).

In studies comparing EUS-*Aphanomyces* isolates with isolates from mycotic granulomatosis (MG) and red spot disease (RSD) outbreaks, Hart (1997) analysed the ITS1-ITS4 region of the rRNA gene cluster using 10 enzymes (*Alu* I, *Dde* I, *Hae* III, *Hha* I, *Hinf* I, *Hpa* II, *Hsp*92 II, *Mbo* I, *Rsa* I and *Sau*96 I); and sequenced the NS5-NS6 and ITS1-ITS2 regions. The isolates had all previously been shown to be slow-growing and pathogenic to snakehead fish when injected intramuscularly (Lilley and Roberts, 1997). These studies found no differences between any of the EUS, MG and RSD isolates.

A variety of saprophytic *Aphanomyces*, *Achlya* and *Saprolegnia* species isolated from the surface of EUS-affected fish or from infected waters, and Oomycete fungi involved in other diseases of aquatic animals, were also included in these analyses. These isolates had previously been shown to have very different temperature-growth profiles from the pathogens, and were incapable of growth within snakehead fish (Lilley and Roberts, 1997).

The rRNA gene studies succeeded in differentiating all of these isolates from the EUS, MG and RSD pathogens (Hart, 1997). Dendrograms constructed from the RFLP data showed that the *Aphanomyces* pathogens clustered most closely to European isolates of the crayfish plague fungus, *Aphanomyces astaci*.

Lilley *et al.* (1997) used RAPD-PCR of genomic DNA to investigate possible intra-specific differences between the isolates. Twenty pathogenic isolates from several localities in Bangladesh, Thailand, Indonesia, Philippines, Australia and Japan were compared using 14 ten-mer primers. Also included in the study were 6 *Aphanomyces* saprophytes, 4 *A. astaci* isolates, and 2 *Aphanomyces* isolates from fish affected by ulcerative mycosis (UM) off the eastern coast of the USA. A total of 321 bands were used for the analysis.

The mean similarity (F ± SD) between all the pathogens was calculated at 0.95 ± 0.03, whereas the other *Aphanomyces* species had a mean similarity of only 0.14 ± 0.05 compared with the pathogens. These results show that the EUS, MG and RSD pathogens are not only con-specific (now listed in the Index of Fungi as *Aphanomyces invadans*), but also genetically very similar. This indicates that the isolates are not long-term residents in each locality, but have spread across Asia relatively recently. In comparison, RAPD studies on *A. astaci* yielded four distinguishable groups from 15 European isolates indicating that there have been several introductions of that fungus to Europe over a number of years (Huang *et al.*, 1994).

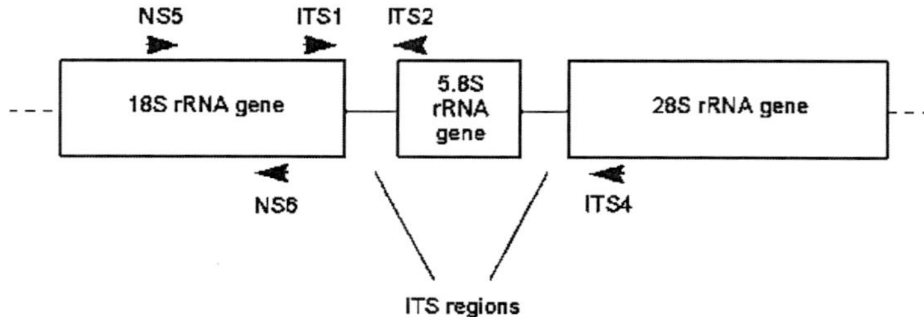

Figure: *Structure of the rRNA gene cluster and positions of fungal PCR primers. The cluster is split into coding (18S, 5.8S and 28S genes) and non-coding (Internally Transcribed Spacer or ITS) regions. The positions of the PCR primers and their direction of synthesis are indicated by arrows.*

Diagnosis of EUS

Ulcerated fish are diagnosed as EUS-positive by histological demonstration of distinctive mycotic granulomas in underlying tissues. This is a reliable technique that yields a lot of information about the disease. As a result of the molecular characterisation work described above, a DNA probe for the specific detection of *Aphanomyces invadans* has been developed, and this could be used in a PCR-based diagnosis of EUS. However, the results of such a test would not give any information on the extent of infection, if indeed the fish is infected and not just carrying propagules of the fungus, or retaining fungal DNA from a past infection.

PCR diagnoses also can suffer from problems of reliability and reproducibility, and in most EUS-affected areas it is a more expensive procedure than histology. Therefore, instead of PCR, attempts are being made to develop an *in situ* hybridisation technique using the probe. It is hoped this will enable histological sections to be further processed for the specific detection of *A. invadans*, and would compliment, rather than replace, histological diagnosis. The development and application of DNA probes for other agents associated with EUS (e.g. rhabdoviruses) would also provide further information about the involvement of these agents in EUS outbreaks.

PCR-Based Method to Detect Aphanomyces Invadans *in the Environment*

To date, isolates of *A. invadans* have only been obtained from internal tissues of EUS-affected fish. Efforts have been made to recover *A. invadans* from natural water bodies in Thailand, but these have not succeeded due to colonisation of isolation media or fungus baits by faster-growing saprophytic fungi (Willoughby and Lilley, 1992). Fraser and Callinan (1996) used particular growth characteristics of *A. invadans* to devise a technique that excludes opportunist fungi. They were able to quantify *A. invadans*-like colonies on two occasions, but it has proved difficult to reproduce this technique reliably. As a result, important aspects of the natural ecology of *A. invadans* (e.g. persistence of the fungus in ponds outwith the EUS season, and fungus viability on resistant/carrier fish or on non-fish substrates) have yet to be studied.

Molecular detection techniques have been used to assay for Oomycete plant pathogens in environmental samples (Judelson and Messenger-Routh, 1996; Coelho *et al.,* 1997; Liew *et al.,* 1998) and DNA probes are presently being developed by researchers studying the toxin-producing dinoflagellate, *Pfiesteria* (Greer *et al.,* 1997). Ulcerative disease outbreaks in eastern USA have been associated with an invasive mycosis as well as with *Pfiesteria* toxins (Blazer *et al.,* 1998). In these cases, molecular techniques may be useful in detecting the various agents and determining their role in each outbreak.

A molecular assay technique has recently been devised to test for the presence of *A. invadans* DNA in water samples and other substrates. This is based on the PCR amplification of an *A. invadans*-specific 98 bp sequence, that was identified during the RFLP work described above. The particular problems of validating a PCR-based proxy detection method were identified by Hiney and Smith (1998) with regards to bacterial fish pathogens. They devised a study framework that evaluated quantitative, qualitative and reliability criteria at 4 levels of experimental complexity: (a) *in vitro*; (b) using a sterile seeded microcosm; (c) in a non-sterile incurred mesocosm; and (d) in non-sterile field samples. This approach can be applied to the study of *A. invadans* propagules in the environment. The planned study levels for this work are listed below.

a) The *in vitro* study aims to assess the specificity and sensitivity (DNA low/high detection limits) of the PCR assay on DNA extracted from cultured fungal mycelium. A range of *A. invadans* isolates from different localities is being studied against a strain panel of related species recovered from EUS-affected areas.
b) Sterile seeded mesocosm. A procedure is being developed to test the DNA extracted from fungal zoospores suspended in flasks of sterile distilled water. *A. invadans* spores will be tested among spores of fungi from the strain panel. Zoospore detection limits will be assessed by making haemocytometre counts of the zoospores present.
c) Non-sterile incurred mesocosm. Fungal zoospore suspensions will be induced in tanks containing pond water. Zoospore detection limits cannot be accurately assessed at this level as fungal species cannot be identified during haemocytometre zoospore counts. It is hoped the tank study can be used to assess the effect of different variables on the persistence of fungal DNA in the water (e.g. water quality, presence/absence of fish and other potential substrates, different treatment regimes).
d) The study of non-sterile field samples aims to test for *A. invadans* DNA in affected areas during non-outbreak periods, and in unaffected areas.

Conclusions

PCR-based techniques have provided valuable information in the study of the fungal pathogen involved in EUS outbreaks. In particular, research on the characterisation of *A. invadans* isolates has benefited from the use of these techniques, and it is envisaged that information on the ecology of the fungus will also be obtained with the use of PCR. It is hoped that the outputs from these latter studies will enable risk factors for the disease to be identified, so that recommendations can be made regarding control methods. It is not envisaged that PCR will be used in the routine diagnosis of EUS, but DNA-based methods may compliment histological diagnoses by providing *A. invadans*-specific stains using *in situ* hybridisation.

6

Role of Irradiation in Fish Preservation

There are three main areas where irradiation is helpful in fish preservation. They are; disinfection, extension of storage life and destruction of pathogens.

Disinfestations

It becomes useful in storing dried fish, by destruction of insects through radiation dose up to 1 KGy. The fish should store appropriately packed to prevent re-infestation. Parasites can also be destroyed by low doses radiation.

Extension of Storage Life

Employing doses up to 2 KGy, the number of food spoiling organisms can be significantly reduced, thus resulting in extension of shelf life.

Destruction of Pathogens

Pathogenic bacteria like *Salmonella* sp and *Listeria* sp can be destroyed with radiation doses between 2 and 5 KGy. These doses are sufficient to destroy only viable cells; higher doses will necessary to eliminate bacterial spores' even frozen foods also.

Safety of Irradited Seafoods

With regard to the irradiated sea foods, considerations of safety for consumption involve four aspects; radiological safety, toxicological safety, microbiological safety and nutritional adequacy.

a) *Radiological safety:* The Joint FAO/IAEA/WHO Expert Committee on the wholesomeness of irradiated food (JECFI) has recommended 10 MeV as the maximum permissible energy for electrons, 5 MeV for X-rays and 10 KGy for gamma rays. The measurement showed that no induced radioactivity could be detected in foods irradiated with 10MeV electrons at 10 KGy. But, when 20 MeV electrons were used, 0.01 Bq/g was detected on the day after irradiation. Radioactivity cannot be induced in foods unless irradiated with radiations above the threshold energy.

b) *Toxicological safety:* The radiation energy absorbed by irradiated foods causes various chemical reactions in the food. The chemical reactions induced by >10 KGy dose lead to about 300mg of radiolytic products per kilogram of food. Some are unique to irradiated foods and of unknown toxicity. Several studies have shown that no toxicological hazard to human health would arise from consumption of food irradiated up to an average dose of 10 KGy. No evidence of toxicological hazards associated with higher doses; however, the recommended dose is up to 10 KGy.

c) *Microbiological safety:* Vegetative cells are most radiation sensitive, although some bacteria viz. *Deinococcus radiodurans* and members of *Moraxella, Actinobacter* require comparatively higher doses for their inactivation. Low (up to 1 KGy) and medium (1-10 KGy) radiation treatments suppress the spoilage causing gram -ve organisms such as *Pseudomonas sp., Proteus sp., Aeromonas sp.,* etc. Resistance to irradiation by certain bacteria such as *Micrococcus radiodurans* and *Micrococcus radiophilus* has been attributed to their unusual ability to repair the breaks in DNA caused by irradiation. Studies have shown that radiation resistance of microorganisms is influenced by treatment conditions such as temperature, presence of air, nitrogen, vacuum, etc. As compared to bacteria, viruses require higher radiation doses for their inactivation. Irradiation process appears little different to other physical processes in its microbiological changes such as mutation leading to increased resistance, enhanced pathogenicity of changed physiological traits important to their identification. In practice, the microbiological risks associated with irradiation are small.

d) *Nutritional adequacy:* The nature and extent of the effects of ionizing radiation on nutrients depend on the composition of food, the radiation dose and modifying factors such as temperature and presence / absence of oxygen. Irradiation up to 10KGy does not significantly alter the nutritional value of proteins, carbohydrates, minerals or saturated fats.

   i) *Proteins: In vitro* studies have shown that free amino of proteins are sensitive to radiation. A large proportion of radiation energy deposited in an irradiated protein leads to protein denaturation, although it is much less compared with heating. Structural changes caused by the radiolytic reaction in food proteins may cause changes in functional properties such as viscosity.

   ii) Lipids: Radiolytic products can cause oxidation of lipids leading to rancidity. Oxidation reactions can lead to the loss of essential unsaturated fatty acids. Ozone, a strong oxidizer is produced from oxygen during food irradiation and may oxidize lipids and also myoglobin resulting in discolouration and flavour changes.

   iii) *Vitamins:* Among the vitamins, thiamine is more radiation sensitive. Other radiation sensitive vitamins are A, E, C and K. The loss of vitamins is insignificant below 1 KGy dosage. However different reports suggest that the vitamin loss caused by irradiation appears to be contradictory.

Organoleptic Changes: Irradiation causes only marginal changes in flavour, texture, and odor of fish. In fatty fishes, the free radicals produced as a result of irradiation can initiate autoxidation chain that will lead to rancidity. The radiation — included flavour changes, described as 'metallic', 'burnt feather — like', 'rubbery' etc. are associated with irradiated fish even when the dosage low. Some other organoleptic changes associated with sterilizing doses are brown discolouration, toughening of texture and strong bitter flavours develop in fish. Some fat — based pigments get bleached during irradiation. Integrating packaging under vacuum with irradiation can suppress bacterial spoilage and oxidative rancidity.

Packaging of Irradiated Fish: Flexible packages have special advantages and many plastic materials can be used in conjunction with irradiated foods. Ionising radiation may also be used to improve the properties of polymeric packaging materials and inactivate any micro organisms, with the packing material might have been contaminated with prior to bringing into contact with the food. In contact with food, extractives from the package should not contaminate the food. The packaging material does not get itself irradiated and induce radioactivity in the food contained.

Legal Aspects and Public Health

Governments all over the world have the responsibility to control irradiation processing of food. The International Consultative Group on Food Irradiation (ICGFI) which was established by the International Organisations such as WHO, FAO, IAEA, has issued guidelines for preparing regulations to control food irradiation facilities. Any new method of preservation has necessarily to ensure that no toxic substances are produced due to the process and the foods preserved by such methods should not exhibit deleterious effects of any sort on the health of the consumer.

Doubts about risk of induced radioactivity in such foods have been expressed at times. No report of formation of harmful toxic substances is available, even though peroxides are formed to some extend in fatty foods during irradiation. No evidence of development of carcinogenicity in such foods has yet been put forth.

The irradiation processing has to be used only as a supplement and should never form substitute for good hygienic practices because it depends upon the initial quality of the raw material. Foods that have been irradiated must be labelled with green international logo to inform that the food has been processed by ionizing radiation. The words "Treated with radiation" or "Treated by irradiation" must be in the same print style as the product name and be no smaller than one-third the size of the largest letter in product name.

Conclusion

Thus, it can be concluded that exposure of fishery products to ionizing radiation could effectively eliminate or reduce the pathogens of public health significance, spoilage causing microorganisms, insects and parasites while maintaining wholesomeness and sensory quality of the seafood. Such technology is more useful for the countries like India to meet the nutritional demands of huge population, as fish is a cheap animal protein source with high essential amino acids as well as with health beneficial omega — 3 polyunsaturated fatty acids.

Aquatic Biodiversity : Threats and Conservation

Biodiversity or Biological Diversity a sum of all the different species of animals, plants, fungi, and microbial organisms living on Earth and the variety of habitats in which they live. Each species is adapted to its unique niche in the environment, from the peaks of mountains to the depths of deep-sea hydrothermal vents, and from polar ice caps to tropical rain forests. According to the definition of the Convention on Biological Diversity, biodiversity is the variability among living organisms from all sources, including terrestrial, marine and other aquatic ecosystems and the ecological complexes of which they are part; this includes diversity within species, between species and of ecosystems.

Aquatic biodiversity can be defined as the variety of life and the ecosystems that make up the freshwater, tidal, and marine regions of the world and their interactions. Aquatic biodiversity encompasses freshwater ecosystems, including lakes, ponds, reservoirs, rivers, streams, groundwater, and wetlands. It also consists of marine ecosystems, including oceans, estuaries, salt marshes, seagrass beds, coral reefs, kelp beds, and mangrove forests. Aquatic biodiversity includes all unique species, their habitats and interaction between them. It consists of phytoplankton, zooplankton, aquatic plants, insects, fish, birds, mammals, and others.

Importance of Aquatic Biodiversity

Aquatic biodiversity has enormous economic and aesthetic value and is largely responsible for maintaining and supporting overall environmental health. Humans have long depended on aquatic resources for food, medicines, and materials as well as for recreational and commercial purposes such as fishing and tourism. Aquatic organisms also rely upon the great diversity of aquatic habitats and resources for food, materials, and breeding grounds. Factors including overexploitation of species, the introduction of exotic species, pollution from urban, industrial, and agricultural areas, as well as habitat loss and alteration through damming and water diversion all contribute to the declining levels of aquatic biodiversity in both freshwater and marine environments. As a result, valuable aquatic resources are becoming increasingly susceptible to both natural and artificial environmental changes. Thus, conservation strategies to protect and conserve aquatic life are necessary to maintain the balance of nature and support the availability of resources for future generations.

Threats to Aquatic Biodiversity

Human activities are causing species to disappear at an alarming rate. Aquatic species are at a higher risk of extinction than mammals and birds. Losses of this magnitude impact the entire ecosystem, depriving valuable resources used to provide food, medicines, and industrial materials to human beings. Runoff from agricultural and urban areas, the invasion of exotic species, and the creation of dams and water diversion have been identified as the greatest challenges to freshwater environments (Allan and Flecker 1993; Scientific American 1997).

Overexploitation of aquatic organisms for various purposes is the greatest threat to marine environments, thus the need for sustainable exploitation has been identified by the Environmental Defense Fund as the key priority in preserving marine biodiversity. Other threats to aquatic biodiversity include urban development and resource-based industries, such as mining and forestry that destroy or reduce natural habitats. In addition, air and water pollution, sedimentation and erosion, and climate change also pose threats to aquatic biodiversity.

1. Overexploitation of species — Overexploitation of species affects the loss of genetic diversity and the loss in the relative species abundance of both individual and /or groups of interacting species. The population size gets reduced because of disturbances in age structure and sex composition. Efficient gears remove quick growing larger individuals . consequently, the proportion of slow growing ones increases and the average size of individuals in a population decreases. Over-fishing causes change in the genetic structure of fish populations due to loss of some alleles. Thus, genetic diversity gets reduced .
2. Habitat modification — Physical modification of habitat may lead to species extinction. This is mainly caused due to damming, deforestation, diversion of water for irrigation and conversion of marshy land and small water bodies for other purposes. Construction of dams on river impedes upstream migration of fishes and displaces populations from their normal spawning grounds and separate the population in two smaller groups. Deforestation leads to catchment area degradation due to soil erosion which results into sedimentation and siltation. This not only affect the breeding ground of aquatic organisms but cause gill clogging of small fishes also.

3. Pollution load — Four forms of pollutants can be distinguished-
   i. Poisonous pollutants — Agrochemicals, metals , acids and phenol cause mortality, if present in a high concentration and affect the reproductive functionality of fish (Kime, 1995).
   ii. Suspended solids — it affects the respiratory processes and secration of protective mucus making the fish susceptible to infection of various pathogens.
   iii. Seewage and organic pollutants — They cause deoxygenation due to eutrophication causing mortality in fishes.
   iv. Thermal pollution — It cause increase in ambient temperature and reduce dissolved oxygen concentration leading to death of some sensitive species.

These factors affect the aquatic biodiversity directly or indirectly. Excessive mortality of organisms due to any of these factors may lead to two type of effects – i) extinction of the species / populations ii) reduction of population size.

Conservation Approaches

Aquatic conservation strategies support sustainable development by protecting biological resources in ways that will preserve habitats and ecosystems. In order for biodiversity conservation to be effective, management measures must be broad based.

- Aquatic areas that have been damaged or suffered habitat loss or degradation can be restored. Even species populations that have suffered a decline can be targeted for restoration (e.g., Pacific Northwest salmon populations).
- An aquatic bio- reserve is a defined space within a water body in which fishing is banned or other restrictions are placed in an effort to protect plants, animals, and habitats, ultimately conserving biodiversity. These bio-reserves can also be used for educational purposes, recreation, and tourism as well as potentially increasing fisheries yields by enhancing the declining fish populations. These bio-reserves are also very similar to marine protected areas, fishery reserves, sanctuaries, and parks.
- Bioregional management is a total ecosystem strategy, which regulates factors affecting aquatic biodiversity by balancing

conservation, economic, and social needs within an area. This consists of both small-scale biosphere reserves and larger reserves.

- Watershed management is an important approach towards aquatic diversity conservation. Rivers and streams, regardless of their condition, often go unprotected since they often pass through more than one political jurisdiction, making it difficult to enforce conservation and management of resources. However, in recent years, the protection of lakes and small portions of watersheds organised by local watershed groups has helped this situation.
- Plantation of trees in the catchment area of water body prevent soil erosion and subsequently reduce the problem of slitation in water body resulting in better survival of aquatic organisms.
- Avoid the establishment of industeries, chemical plants and thermal power plants near the water resources as their discharge affect the ecology of water body resulted in loss of biodiversity.
- The World Resources Institute documents that the designation of a particular species as threatened or endangered has historically been the primary method of protecting the biodiversity.
- Many specialized programs should be instituted to protect biodiversity. For example, the USDA Forest Service started a cooperative state-federal program with a goal to restore the health of riverine systems and associated species.
- Regulatory measures must be taken on wastewater discharge in the water body to conserve biological diversity.
- Increasing public awareness is one of the most important ways to conserve aquatic biodiversity. This can be accomplished through educational programs, incentive programs, and volunteer monitoring programs.
- Various organisations and conferences that research biodiversity and associated conservation strategies help to identify areas of future research, analyze current trends in aquatic biodiversity.

The Role of Immunostimulants in Indian Aquaculture

Aquaculture is the fastest growing food production sector in the world. Among the Asian countries, India ranks second in aquaculture and third in capture fisheries. Over the last few decades, movement of animals and animal products such as broodstock, seed and feed, emergence of new diseases and their potential establishment in new area, the irresponsible use of chemical disinfectants and antibiotics are recognised as having potential impacts on environment.

Disease is major limited factor in fish culture. Epizootic ulcerative syndrome (EUS) in 1986 and White spot diseases in 1994 and other diseases like vibriosis, fin and tail rot, dropsy have had divesting impact on aquaculture production in India and Asia. An infectious disease is a major contributor to economic loose in intensive fish culture.

Farmers of India have still following semi-intensive culture system where the chances of fish in stress is more due to overcrowding and other environmental factors. Stress weakens fishes' immune systems, leading to increased susceptibility to disease. Aquaculture faces serious problems due to various adverse effects of antibiotics such as accumulation in the tissue and immunosuppression. Moreover, due to the availability of limited vaccines in few countries and their pathogen specific protective action, much attention has been directed towards the use of immunostimulants in aquaculture to control infectious diseases.

Immunostimulation is one of the useful tools in aquaculture where vaccination and/or treatment by injections are difficult and labourious processes, and where repeated chemotherapy poses a problem of developing drug resistance strains of pathogens. Immunostimulants potentiate the immunity of the host itself, enabling it to defend more strongly against pathogens.

Definition of Immunostimulants

An Immunostimulant is a chemical, drug, stressor or action that elevates the non-specific defence mechanism or the specific immune response. Immunostimulants may be given by themselves to activate non-specific defence mechanism or may be administered along with a vaccine to activate non-specific mechanism as well as heightening a specific immune response.

List of pathogen successfully controlled by immunostim-ulants exposure in fish/shrimp:

Bacteria: Aeromonas hydrophila, A. salmonicida, Edwardsiella tarda, E. ictaluri, Vibrio anguillarum, V.vulnificus , V. salmonicida, Yersinia ruckeri, Streptococcus sp.

Virus: Infectious hematopoietic necrosis, yellow head virus, viral hemorrhagic septicemia.

Parasite: Ichthyopthirius multifiliis.

Immunostimulants Used in Fish and Shrimp

Bacterial Derivatives: Some bacterial component stimulate cellular and inflammatory responses in animals; logically so, as the animal are genetically and environmentally conditioned to combat these pathogens. Indeed, if the toxicity and harmful inflammatory factors are deleted, bacterial derivatives can be good Immunostimulants. For instance muramyl dipeptide (MDP), the mycobacterium peptidoglycan derivatives has been used experimentally in clinical use for immunostimulating response in cancer patients. Another Immunostimulant FK-565, a synthetic lactoyl tetrapeptide based on a component from Streptomyces olivaceogriseus sp. is known to induce stimulation against Aeromonas salmonicida in rainbow trout. Besides elevated phagocytosis, enhanced intracellular killing also observed.

Lipopolysaccarides (LPS) preparation from gram negative bacteria is used as Immunostimulant. LPS preparations have been tested in fish and are known to stimulate B cell proliferation if given in vivo or in vitro in appropriate doses.

These substances are very potent even in very low doses and may occur as contaminants in bacterin preparations thus some bacterin used in fish immunizing programs may have their own Immunostimulants.

Yeast Derivatives: Glucans, long chain polysaccharide extracted from yeast, are good stimulators of non specific defence mechanism in animals including fish and shellfish. Cultured fish and shrimp treated with these glucan have showed increased phagocytic activity and protection against some bacterial pathogens. Glucans have tremendous potential as additives in fish food for developing broad spectrum protection against bacterial disease.

Algal Derivatives: Laminaran is a B (1, 6)-branched B (1, 3)-D-glucan, a major component in sublittoral brown algae, e.g: Phaeophyceae. Almost all B-(1, 3) D-glucan display poor water solubility which makes them less easy to handle than aqueous soluble laminaria. Laminaran obtained from Laminaria hyperhorea has immunomodulatory effect on anterior kidney. Interperitonial injection of laminaran has also been shown to be preventing mortality caused by Aeromonas hydrophila injection in Blue gourami.

Synthetic compounds:-

Muramyl peptides

Levamisole

FK-565

Animal and plant extracts:-

Chitin and Chitosan

Chitin is a polysaccharide forming the principle component of crustacean and insect exoskeleton and the cell walls of certain fungi. Rainbow trout injected with chitin showed stimulated macrophage activities resistant to Vibrio anguillarum infection.

Chitosan, de-*N*-acetylated chitin can increase in immunological parameters in blood such as NBT, potential killing activity, myeloperoxidase and total immunoglobulin concentration and give resistance from certain bacteria.

Vitamins: Vitamin C (ascorbic acid) has influence on fish macrophage function such as engulfment and destruction of bacteria. Vitamin C also neutrophils serving to inactive oxygen radicals to protect the cell from injury or promoting the oxidative destruction of bacteria. Enhanced antibody production and complement activity in channel catfish and rainbow trout due to vitamin c has been demonstrated. The role of vitamin c to enhance inflammatory response and disease resistance has been recently demonstrated in Indian major carps. Vitamin E also has shown to influence non-specific and specific defense mechanism in rainbow trout. It also plays an important role in cell membrane structure, stability and function.

Hormones: The relationship between neuroendocrine regulation and the immune system has recently become the subject of intense investigation. It is also known that growth hormone (GH) and Prolactin directly affect the immune competent cells (macrophage, lymphocytes....). Melanin stimulating hormone (MSH) and melanin concentrating hormone (MCH) stimulate phagocytosis by head kidney leucocytes of rainbow trout in vitro.

Time of Administration of Immunostimulant

Immunostimulants may be given separately or along with a vaccine. The fish could be prepared for predicted events, such as seasonal exposure to a pathogen by a treatment prior to event. Easiest time for administration of immunostimulant and vaccine is at the same time as exposure of fish to the specific antigen.

The two substances are mixed with feed or delivered through immersion. In some cases Immunostimulant is better given prior to vaccine thus helping prepare fish for antigen exposure. However, dosage and timing should for each substance should be determined.

Efficacy of Immunostimulant

Lower temperature has profound influence on the effect of Immunostimulant. So also each species, depending on its evolutionary divergence respond differently to Immunostimulant. Higher doses of Immunostimulant suppress the immune responses. Duration of administration is also important. Additionally, fish are subjected to daily and seasonal cycles that may affect the efficacy and potency of Immunostimulant.

Status of Immunostimulants in India

A variety of natural and synthetic substances have been tried in Indian major carps and shrimp in India. B-1-3 glucan, levamisole, Chitosan, ascorbic acid, tuftsin, á-tocophrol and quartenary ammonium compound have been tried In Indian conditions and show immunostimulating effect on various fish.

Conclusion

The use of immunostimulants has opened a new chapter and a very promising area in aquaculture. Immunostimulants may be used to prevent looses from disease. Although vaccination is the most reliable method to control fish diseases, as yet, no effective vaccine against Bacterial Kidney disease or most viral infections and there is no single commercial vaccine available in India, and imported vaccine may not be effective because the strain which cause disease may be different in India. Immunostimulants are safer than chemotherapy and their range of efficacy is wider than vaccination. Therefore immuno-stimulants may be a powerful tool for controlling infectious diseases.

Broodstock Nutrition and Management in Crustacean

The future of aquaculture production, as in other animal production systems, is towards greater control of the physical, chemical and biological variables surrounding the production system. The trend from collection of wild fry to the development of hatcheries for fry production, from the use of wild spawners to the use of maturation systems and ultimately closing the reproductive cycle to use domesticated and selected stocks is clear. At the same time, there is an increasing trend towards production efficiency and cost-effectiveness, not least in the hatchery stage, particularly where the supply — demand situation progresses from quantity production to quality production.

These trends are becoming more evident in all forms of commercial aquaculture. Broodstock and larval nutrition are key elements underpinning this progress towards greater control and domestication. However, there is substantial work to be done if rearing of aquaculture species is to approach the level of control and understanding which are evident in the poultry, swine and ruminant sectors. Research into shrimp broodstock nutrition is gaining importance with the increasing use of domesticated and genetically selected stocks for aquaculture.

Also, given the high cost of feed in a captive broodstock maturation facility, estimated at around 50% of total cost, as well as a heavy reliance on fresh feed items such as squid, mollusk meals and polychaete worms which may vary in quality and availability, there is a demand for feeds specifically formulated for use in maturation diets. It has also been pointed out (Doyle, personal communication) that the development of genetically improved lines of shrimp may benefit from a simultaneous development of feeds specifically tailored to individual strain requirements.

During penaeid shrimp maturation, nutrient reserves, mainly from the hepatopancreas, are mobilised to support ovarian and testicular maturation, gametogenesis and vitellogenesis (Harrison, 1990; 1997). Tissue reserves in the hepatopancreas can be depleted rapidly so that the diet becomes the most important contributor of nutrients to the developing egg.

This is particularly true when, as is the case in most captive maturation facilities, eyestalk ablation is used to accelerate the process of gonadal maturation. The hormonal and metabolic changes that come around during such forced maturation may take place when the nutrient reserves are insufficient to support rapid ovarian development placing an even larger burden on the diet as a source of essential nutrients. The nutrition of penaeid shrimp broodstock has recently been reviewed (Wouters et al., 2001a). The authors recognise that there is a dearth of information in the scientific literature on the subject, possibly due to the expense and complexity of running sufficiently rigorous nutrition experiments on shrimp broodstock. In addition, much of the research has been carried out with fresh feeds, either alone or in combination with formulated diets or dietary supplements.

Lipids

Due to the importance of lipids in crustacean maturation, much of the work carried out to date has focused on this aspect, particularly the requirement for highly unsaturated fatty acids (HUFA) and phospholipid. During maturation, lipids are mobilized from the hepatopancreas in many species and dietary lipids rapidly processed for transport to the developing ovaries. Total lipid does not appear to be important although Wouters et al. (2001b) reported that excessively high total lipid in the diet had an adverse effect on ovarian maturation and feed consumption, possibly due to satiation. However, the average lipid level in commercial broodstock diets (10%) appears to be around 3% higher than in grower feeds used in commercial culture ponds.

Highly unsaturated fatty acids (HUFA), especially 20:5n-3 and 22:6n-3, are abundant in ovarian tissues and are believed to be an important component of live and formulated maturation diets. Diets deficient in n-3 HUFA have been found to have a negative effect on ovarian development, fecundity and egg quality (Wouters et al., 1999a). Arachidonic acid (20:4n-6; AA) has been detected at high levels in the ovaries of wild shrimp and is also abundant in some of the best fresh feed items such as polychaetes (bloodworms), clams and mussels (Harrison, 1997). The n-6 HUFA are known to be precursors of the prostaglandin hormones, which act in reproduction and vitellogenesis.

According to Wouters et al. (2001a), formulated maturation diets appear to be deficient in AA as well as relatively low eicosapentanoic acid (EPA) levels. It has been proposed that the ratio of n-3 to n-6 levels in the diet is important and that it should be around 2—3 to 1 (Wouters et al., 1999b).

Phospholipids, mainly phosphatidylcholine and phosphatidylethanolamine appear to be predominant in the shrimp ovary and there seems to be a requirement for phospholipids in the diet. Several studies (Ravid et al., 1999; Wouters et al., 1999b) have demonstrated the effects of phospholipid levels in the diet and it has been suggested that broodstock diets should contain more than 2% phospholipid to ensure that 50% of total egg lipid is in this form (Cahu et al., 1994).

Cholesterol is the precursor of steroid hormones and it is known that shrimp have a requirement for cholesterol in the diet. Cholesterol is stored in the hepatopancreas and is mobilized during maturation. Some of the live feed organisms used in maturation diets have relatively high cholesterol levels (e.g. squid, clams) although to date there has been limited research into the effects of dietary cholesterol on maturation and reproduction (Wouters, 2001). During maturation, the level of triacylglycerides (TAG) in the ovaries increases as they are incorporated into the egg and decrease after spawning (Ravid et al., 1999; Wouters et al., 1999b). Triacylglycerides appear to be the principal energy source in eggs and nauplii and their importance in reproduction, and egg and postlarval quality has been shown (Palacios et al., 1998; 1999).

Protein and Amino Acids

Maturation is a time of intense protein synthesis and it is likely that the requirement for protein is higher at this time (Harrison, 1997). Wouters et al. (2001a) report that the protein content of formulated diets in their studies was around 50% but that this was still low compared to the level in fresh feeds.

Some studies have shown changes in protein content of the ovaries associated with egg development and spawning, and with spawning success. Harrison (1997), found an increase in ovarian protein levels associated with ovarian development followed by a sharp decrease after spawning in the shrimp Paratelphysa hydrodromaus and several authors have noted a similar increase in farmed penaeid shrimp (Castille and Lawrence, 1989).

A marked difference has also been noted in the protein content of the hepatopancreas and ovaries of wild and domesticated females of Litopenaeus vannamei with good repeat spawning performance which have been found to have significantly higher protein content than females with poorer spawning performance (Palacios et al., 2000).

Carbohydrates

Carbohydrates do not appear to be essential for shrimp broodstock diets although Palacios et al. (1998; 1999) related egg glucose levels with larval quality and broodstock condition. Carbohydrates can be used as cost-effective ingredients to contribute to glycogen accumulation in the hepatopancreas as well as providing other benefits in the broodstock diet, acting as binders and possibly playing a role in transport of nutrients in the hemolymph (Harrison, 1997).

Vitamins and Minerals

Detailed vitamin and mineral requirements for shrimp broodstock diets are relatively unknown with only a few studies on vitamins A, C and E. Alava et al. (1993) found that ovarian maturation was slower when fed a diet deficient in either vitamins E, A and C. Vitamin E appears to be important in crustacean broodstock nutrition. Chamberlain (1988), reported a correlation between vitamin E deficiency and the percentage of abnormal sperm in Litopenaeus setiferus and Cahu et al. (1991) found an improvement in hatching rate with increasing dietary vitamin E correlated to increasing—tocopherol levels in the egg. Wouters et al. (1999b) found a similar correlation to that observed by Cahu et al. (1991) between spawn and hatch quality with a-tocopherol levels in wild spawners and nauplii of L. vannamei. They found that mature ovaries and nauplii contained higher levels of —tocopherol than immature and spent ovaries. Harrison (1997) also speculated that vitamin E in the egg yolk may also act as a natural antioxidant. Harrison (1997), suggested the importance of dietary vitamin A due to its accumulation in the ovaries of crustaceans during maturation. Vitamin C content of eggs of Fenneropenaeus indicus are also influenced by the levels in the diet, and high hatching rate has been related to high ascorbic acid levels in the eggs (Cahu et al., 1995). Harrison (1997) assumes that vitamin D is important in broodstock diets due to its probable role in calcium and phosphorus metabolism in crustaceans. Harrison (1990) discussed the possibility that mineral deficiencies or imbalances could have a negative impact on crustacean reproduction and may play a role in oocyte resorption, reduction in reproductive performance and egg quality. Studies into mineral requirements are rare due to several complications (Wouters et al., 2001a). Where studies have been conducted, diets were formulated with mineral mixes with added calcium, phosphorus, magnesium, sodium, iron, manganese and selenium (Marsden et al., 1997; Xu et al., 1994).

Spent broodstock of L. vannamei had lower levels of calcium and magnesium in the muscle and lower magnesium levels in the hepatopancreas (Mendez et al., 1997), possibly due to a combination of dietary deficiencies and losses through moulting and transfer to the eggs. Copper also decreased in the hepatopancreas, possibly through transfer to the ovaries, although it increased in the muscle tissue. It is clear that more studies need to be undertaken into mineral nutrition in broodstock diets.

Carotenoids

Crustaceans cannot synthesise carotenoids de novo, and a dietary source of these pigments is required. During sexual maturation, most crustaceans accumulate carotenoids in the hepatopancreas and during vitellogenesis, these are transported in the hemolymph as carotenoglycolipoproteins to accumulate in the eggs as part of the lipovitellin protein. Dall (1995) found that free astaxanthin levels in the developing ovaries of Penaeus esculentus increased from 2 to 34 ppm and in the digestive gland, from 20 to 120 ppm. Carotenoids, especially astaxanthin, are strong antioxidants and probably play a role in protecting the broodstock nutrient reserves and developing embryos from oxidation (Dall et al., 1995; Merchie et al., 1998). It has also been suggested (Harrison, 1997) that they act as pigment reserves in the embryos and larvae for the development of chromatophores and eyespots, and as a vitamin A precursor (Dall, 1995). Pangantihon-Kühlmann and Hunter (1999) found that astaxanthin supplementation (50 mg/kg) of the diet resulted in increased egg production in Penaeusmonodon but could not demonstrate any benefit of astaxanthin supplementation on either hatching rate or metamorphosis to zoea 1 stage.

Miscellaneous Factors

Hormones

It has been suggested that some of the more successful live feed organisms may provide benefits through the provision of hormones or their precursors. Naessens et al. (1997) speculated that part of the reason for the success of reproductive adult Artemia biomass supplementation of the diet of L. vannamei broodstock could be due to the presence of specific hormones or analogous peptides in the Artemia that provoked a response in the shrimp. Bloodworms used in maturation have also been found to contain methyl farnesoate, an ecdysone hormone that has been shown to increase reproductive performance in the spider crab Libinia emarginata (Laufer et al., 1987), L. vannamei (Laufer et al., 1997), P. monodon (Hall et al., 1999) and the crayfish Procambarus clarkii (Laufer et al., 1998). In P.

clarkii, the hemolymph titer increased from basal levels during early vitellogenesis, peaked during mid-cycle and then returned to basal levels when the ovaries were in late vitellogenesis.

Nucleotides

Nucleotides, the basic building blocks of nucleic acids, are recognised as important elements in mammalian nutrition especially during periods of rapid growth or physiological stress (Van Buren, 1994) and also appear to play a key role in the immune system. Traditionally, nucleotides have not been considered essential nutrients although de novo synthesis and salvage pathways are thought to be costly processes in metabolic terms. Several studies have demonstrated that dietary sources of nucleotides can have beneficial effects and the term 'conditionally essential' has been used to describe their role in nutrition (Carver and Walker, 1995). Exogenous sources of nucleotides are thought to optimise the functions of rapidly dividing tissues, such as those of the developing embryo and young, and the reproductive and immune systems. Most aquaculture diet ingredients of animal and plant origin contain nucleotides although there are differences in the concentration and availability. Nucleotide content is particularly high in ingredients such as fish solubles, animal protein solubles, fishmeal, legumes (adenine is particularly high in blackeyed peas), yeast extracts and unicellular organisms such as yeasts and bacteria that are rich in RNA or DNA (Carver and Walker, 1995; Devresse, 2000).

Devresse (2000) noted that the low digestibility of whole yeast compared to yeast extract may be related to the protein (nitrogen) solubility as yeast extract has much higher protein solubility than whole yeast. He also noted that, although fish solubles are highly digestible, they leach easily in water affecting availability. Reproduction and egg development have a high requir-ement for RNA and DNA and it may be expected that increasing the availability of nucleotides in broodstock diets may have a beneficial effect on egg development. Recently, research has demonstrated the effect of a nucleotide-enriched diet for broodstock nutrition in aquaculture (Gonzalez-Vecino, 2002; Gonzalez-Vecino et al., 2003).

Nucleotide enrichment of broodstock diets for Atlantic halibut (Hippoglossus hippoglossus) and haddock (Melanogra-mmus aeglefinus) resulted in a general trend towards better spawning performance and egg quality with the nucleotide diet. Total egg yield was 30% higher in the halibut fed with the nucleotide diet and the relative fecundity, mean egg density, hatching rate and survival of yolk-sac larvae were also signi-ficantly improved. Haddock fed on the nucleotide-enriched diet also had significantly higher fertilization and hatching rates. To date, no work has been published on nucleotide supplemen-tation of broodstock diets for shrimp but it would be interesting to conduct some trials to determine if broodstock diets enriched with nucleotides might offer similar benefits in shrimp maturation and breeding. Similarly, the potential for nucleotide supplementation of diets for shrimp larvae should also be investigated.

Conclusion

There are a fewer information regarding the knowledge of bloodstock and larval nutrition of crustaceans available which creates many gaps between researcher and literature . This is partly the result of the complexity and expense of conducting research into these two areas. However, given the increasing importance of domestication in shrimp aquaculture particularly, there is a need for an increased focus on these areas. The role of nutrition in broodstock and maturation performance and in increasing larval survival and quality will be fundamental to obtaining optimal performance from domesticated stocks. Even in species where domestication is not an issue, the improvement of maturation performance and larval production remains a key goal in improving the efficiency of production systems. It is not clear how far the goal of complete replacement of live feeds may be. It is likely that this will be attained in broodstock diets long before larval feeds. Indeed, given the complexities inherent in supplying a complete nutritional package in a small particle, it may be that this goal is never reached.

However, it may be possible to supply a range of diet particles (e.g. high lipid particles, high protein particles, carbohydrate particles etc.) that are aimed specifically at providing the right mix of nutritional elements in the culture tank that will expose the larvae to the appropriate nutritional mix. Alternatively, it may be that, as expressed by D'Abramo (2002), the complexity of the ontogeny of larval nutritional physiology may mean that technical success will be based on a compromise between the desire to provide a complete diet package and the need to strive for simplicity in formulation and manufacture. In this paper maximum emphasis is given on different sources as nutrition and their beneficial effects on broodstock quality but very limited information are available regarding this module. This information will provide a base line for researcher for further studies.

7

Functional Genomics

Functional genomics is a field of molecular biology that attempts to make use of the vast wealth of data produced by genomic projects (such as genome sequencing projects) to describe gene (and protein) functions and interactions. Unlike genomics, functional genomics focuses on the dynamic aspects such as gene transcription, translation, and protein–protein interactions, as opposed to the static aspects of the genomic information such as DNA sequence or structures. Functional genomics attempts to answer questions about the function of DNA at the levels of genes, RNA transcripts, and protein products. A key characteristic of functional genomics studies is their genome-wide approach to these questions, generally involving high-throughput methods rather than a more traditional "gene-by-gene" approach.

Goals of Functional Genomics

The goal of functional genomics is to understand the relationship between an organism's genome and its phenotype. The term functional genomics is often used broadly to refer to the many possible approaches to understanding the properties and function of the entirety of an organism's genes and gene products. This definition is somewhat variable; Gibson and Muse define it as "approaches under development to ascertain the biochemical, cellular, and/or physiological properties of each and every gene product", while Pevsner includes the study of nongenic elements in his definition: "the genome-wide study of the function of DNA (including genes and nongenic elements), as well as the nucleic acid and protein products encoded by DNA". Functional genomics involves studies of natural variation in genes, RNA, and proteins over time (such as an organism's development) or space (such as its body regions), as well as studies of natural or experimental functional disruptions affecting genes, chromosomes, RNA, or proteins. The promise of functional genomics is to expand and synthesize genomic and proteomic knowledge into an understanding of the dynamic properties of an organism at cellular and/or organismal levels.

This would provide a more complete picture of how biological function arises from the information encoded in an organism's genome. The possibility of understanding how a particular mutation leads to a given phenotype has important implications for human genetic diseases, as answering these questions could point scientists in the direction of a treatment or cure.

Techniques and Applications

Functional genomics includes function-related aspects of the genome itself such as mutation and polymorphism (such as SNP) analysis, as well as measurement of molecular activities. The latter comprise a number of "-omics" such as transcriptomics (gene expression), proteomics (protein expression), and metabolomics. Functional genomics uses mostly multiplex techniques to measure the abundance of many or all gene products such as mRNAs or proteins within a biological sample. Together these measurement modalities endeavour to quantitate the various biological processes and improve our understanding of gene and protein functions and interactions.

At the DNA Level

Genetic Interaction Mapping

Systematic pairwise deletion of genes or inhibition of gene expression can be used to identify genes with related function, even if they do not interact physically. Epistasis refers to the fact that effects for two different gene knockouts may not be additive; that is, the phenotype that results when two genes are inhibited may be different from the sum of the effects of single knockouts.

The ENCODE Project

The ENCODE (Encyclopedia of DNA elements) project is an in-depth analysis of the human genome whose goal is to identify all the functional elements of genomic DNA, in both coding and noncoding regions. To this point, only the pilot phase of the study has been completed, involving hundreds of assays performed on 44 regions of known or unknown function comprising 1% of the human genome.

Important results include evidence from genomic tiling arrays that most nucleotides are transcribed as coding transcripts, noncoding RNAs, or random transcripts, the discovery of additional transcriptional regulatory sites, further elucidation of chromatin-modifying mechanisms.

At the RNA Level: Transcriptome Profiling

Microarrays

Microarrays measure the amount of mRNA in a sample that corresponds to a given gene or probe DNA sequence. Probe sequences are immobilized on a solid surface and allowed to hybridize with fluorescently labelled "target" mRNA. The intensity of fluorescence of a spot is proportional to the amount of target sequence that has hybridized to that spot, and therefore to the abundance of that mRNA sequence in the sample. Microarrays allow for identification of candidate genes involved in a given process based on variation between transcript levels for different conditions and shared expression patterns with genes of known function.

SAGE

SAGE (Serial analysis of gene expression) is an alternate method of gene expression analysis based on RNA sequencing rather than hybridization. SAGE relies on the sequencing of 10–17 base pair tags which are unique to each gene. These tags are produced from poly-A mRNA and ligated end-to-end before sequencing. SAGE gives an unbiased measurement of the number of transcripts per cell, since it does not depend on prior knowledge of what transcripts to study (as microarrays do).

At the Protein Level: Protein–Protein Interactions

Yeast Two-Hybrid System

A yeast two-hybrid (Y2H) screen tests a "bait" protein against many potential interacting proteins ("prey") to identify physical protein–protein interactions. This system is based on a transcription factor, originally GAL4, whose separate DNA-binding and transcription activation domains are both required in order for the protein to cause transcription of a reporter gene. In a Y2H screen, the "bait" protein is fused to the binding domain of GAL4, and a library of potential "prey" (interacting) proteins is recombinantly expressed in a vector with the activation domain.

In vivo interaction of bait and prey proteins in a yeast cell brings the activation and binding domains of GAL4 close enough together to result in expression of a reporter gene. It is also possible to systematically test a library of bait proteins against a library of prey proteins to identify all possible interactions in a cell.

AP/MS

Affinity purification and mass spectrometry (AP/MS) is able to identify proteins that interact with one another in complexes. Complexes of proteins are allowed to form around a particular "bait" protein. The bait protein is identified using an antibody or a recombinant tag which allows it to be extracted along with any proteins that have formed a complex with it. The proteins are then digested into short peptide fragments and mass spectrometry is used to identify the proteins based on the mass-to-charge ratios of those fragments.

Loss-of-Function Techniques

Mutagenesis

Gene function can be investigated by systematically "knocking out" genes one by one. This is done by either deletion or disruption of function (such as by insertional mutagenesis) and the resulting organisms are screened for phenotypes that provide clues to the function of the disrupted gene.

RNAi

RNA interference (RNAi) methods can be used to transiently silence or knock down gene expression using ~20 base-pair double-stranded RNA typically delivered by transfection of synthetic ~20-mer short-interfering RNA molecules (siRNAs) or by virally encoded short-hairpin RNAs (shRNAs). RNAi screens, typically performed in cell culture-based assays or experimental organisms (such as C. elegans) can be used to systematically disrupt nearly every gene in a genome or subsets of genes (sub-genomes); possible functions of disrupted genes can be assigned based on observed phenotypes.

Functional Annotations for Genes

Genome Annotation

Putative genes can be identified by scanning a genome for regions likely to encode proteins, based on characteristics such as long open reading frames, transcriptional initiation sequences, and polyadenylation sites. A sequence identified as a putative gene must be confirmed by further evidence, such as similarity to cDNA or EST sequences from the same organism, similarity of the predicted protein sequence to known proteins, association with promoter sequences, or evidence that mutating the sequence produces an observable phenotype.

Rosetta Stone Approach

The Rosetta stone approach is a computation method of de novo protein function prediction, based on the hypothesis that some proteins involved in a given physiological process may exist as two separate genes in one organism and as a single gene in another. Genomes are scanned for sequences that are independent in one organism and in a single open reading frame in another. If two genes have fused, it is predicted that they have similar biological functions that make such coregulation advantageous.

Functional Genomics and Bioinformatics

Because of the large quantity of data produced by these techniques and the desire to find biologically meaningful patterns, bioinformatics is crucial to analysis of functional genomics data. Examples of techniques in this class are data clustering or principal component analysis for unsupervised machine learning (class detection) as well as artificial neural networks or support vector machines for supervised machine learning (class prediction, classification).

Nutrigenomics in Aquaculture Research: A Key in the 'Aquanomic' Revolution

Publication of the human genome draft sequence (Lander et al., 2001; Venter et al., 2001), and those of other species represent major milestones in scientific discovery. This genomic unraveling has helped identify the functions of numerous unknown genes and, in certain cases, has assisted in determining the roles of specific genes in disease processes. The genomic revolution incorporates much more than the decoding of organismal genomes alone, however. Critical to the success of this field have been like innovations in molecular and cellular biology, information technology, analytical sciences and in automation (Witkamp, 2005).

A genome sequence does not itself reveal the functionality of its gene, but rather is a portal to understanding gene and gene product roles, such as expression patterns and locations, protein synthesis and how these processes are modified and regulated. Nor does the genome sequence inform us about the functional processes of a gene or protein, such as their involvement in signal transduction pathways and engagement in transcription, mRNA processing, mRNA stability, translation, or post-translational modifications.

Indeed these tasks are left to the biologist, who must decode and map the active genes, and determine the aforementioned functions. Comprehending this complex genomic blueprint demands a holistic approach. The last decade has witnessed a rapid expansion in the field of functional genomics that represents the confederacy of genomics, proteomics, genotyping, transcriptomics and metabolomics.

The huge data sets generated by the 'omics' fields has energized the field of bioinformatics, which has in turn developed new methods by which to acquire, store, share, analyze, present, and manage 'omic'-derived information, allowing the processing and integration of complex and often dissimilar data sets into seamless and coherent searchable databases (Brazma, 2001; Desiere et al., 2002; van Ommen and Stierum, 2002). These approaches provide a powerful means by which numerous and sometimes negligible changes in a genome can be monitored without the need of prior knowledge of a specific mechanism.

Nutrigenomics

In recognising that nutrients: 1) modify gene expression, 2) may alter normal metabolism, and 3) affect health status (Corthésy-Theulaz et al., 2005) the scientific community has rallied to define inter-relationships between diet, health and disease processes using molecular techniques (Kaput and Rodriguez, 2004; Davis and Hord, 2005).

This subdiscipline of functional genomics, termed nutritional genomics or nutrigenomics, endeavours to resolve the influence of dietary chemicals upon the genome and to increase our understanding of how dietary constituents influence metabolism (Müller and Kersten, 2003; Mutch et al., 2005). Nutrigenomics employs a wide portfolio of 'omic' methods because no single technique is suitable for analyzing all different types of molecule or outcomes.

However, of the 50 or so 'omic' terms coined (Mutch et al., 2005), nutrigenomics truely encompasses only four (Table 1), namely transcriptomics, or microarray technologies, which monitors altered mRNA levels for the entire genome (Scheel et al., 2002); proteomics, which encompasses protein structure determination, expression and molecular interactions (Kussmann et al., 2005); metabolomics, which examines changes in metabolites involved in primary and intermediary metabolism (German et al., 2004); and epigenomics, which ascertains DNA methylation patterns, imprinting and DNA packaging (Beck et al., 1999).

Table: *Some 'omics' defined.*

Term	***Definiton***
Nutrigenomics	The study of genome-wide effects of diet or componts thereof on the transcriptome, proteome and metabolome of cells. tissues or organisms at a specifice moment in time.
Genome	The entirn complement of genetic material of an organism
Transcriptomics	The monitoring of the complete set of RNA transeripts produced by the genome at any given time.
Proteomics	The examination of proteome-the complete set of proteins in a cell or tissue-at a specific moment in time. Proteomics attempts to determine the role of specific proteins and how they intereact with other molecules.
Metabolomics	The identification and quantification large sets of metablites from cells or biological fluids and how these may change follwing physilogical disturbance.
Epigenomics	The detection and examination of DNA methylation patterns both spatially and temporally.

Gene Expression Profiling

It is now widely appreciated that many metabolic processes are dependent upon an exquisite synchronization of events among several organs, involving sometimes many thousands of genes (Liu-Stratton et al., 2004; Kato et al., 2005). This complexity is enhanced by the involvement of other molecules that include receptors, hormones and enzymes. Given the aforementioned complexity, and because diets are heterogenous mixtures of chemicals, it is obvious that evaluating the effects of a nutrient, for example, upon health, requires examining more than a single biomolecule or biomarker. The need to access a suite of markers when examining the effect of a nutrient upon a subject is made more intricate because most genes have small sequence differences or polymorphisms and these distinctions may affect proteinprotein or protein-substrate interactions.

Inconsistencies in some epidemiological studies in humans may be explained partly by polymorphisms, as exemplified by those upon folate metabolism and altered risk to colon cancer and responsiveness to hypolipemiant drugs (Davis and Hord, 2005; Ruano et al., 2005). These recognised deviations from the norm in people have led to the development of the concept of 'personalized medicine', wherein microarray gene expression profiling (GEP) is used to provide elite treatment regimens based upon molecular classification of gene subtypes (Chin et al., 2004; Jain, 2004) - molecular diagnoses and medical interventions that will no doubt be refined further by miniaturization with the advent of nanotechnology.

In nutrition research GEP may be used for three distinct purposes (Müller and Kersten, 2003):

1) To assist in the identification and characterisation of basic molecular pathways that may be impacted, either positively or negatively, by nutrients;
2) To provide insights upon specific mechanisms that trigger such beneficial or negative effects and;
3) To identify specific genes altered by nutrients that might prove valuable as molecular biomarkers or nutrient sensors and in gene discovery.

All three areas of application have already provided high bounty especially in the clinical and pharmacological sciences and all indications suggest that this will continue unabated (Müller and Kersten, 2003; Nugent, 2004; Kaput and Rodriguez, 2004; van Ommen, 2004; Chin et al., 2004; Witkamp, 2005; Mutch et al., 2005; Gibney et al., 2005). In contrast to the successes achieved in the biomedical arena however, applying the same biotechnologies to the animal production sector has yet to provide like rewards. Clearly, fundamental differences exist between clinical and animal production settings.

These distinctions will drive the specific applications of functional genomic technologies in the agribusiness sector, especially with regard to meat and egg production, milk quality and even in developing models for xenotransplantation (Rothschild, 2004; Burt, 2005; Womack, 2005). As in terrestrial animal production, aquaculture too will concentrate research in areas that are most likely to provide economic advantage. The following will centre upon the application of GEP in aquaculture research.

Aquanomics

In functional genomic investigations in aquaculture, or 'aquanomics' research, the in vivo model will represent the gold standard in studies designed to evaluate the effects of particular dietary component(s) upon production characteristics. This introduces a number of complications that must be addressed during experimental design.

Species selection will be necessarily contingent upon the availability of specific homologous microarray systems. This in itself becomes problematic since although many thousands of expressed sequence tags are available for carp, catfish, trout and salmon, complete genome sequences are only offered for zebrafish, tetradon or puffer fish, fugu and medaka (Crollius and Weissenbach, 2005).

The development of Atlantic salmon microarrays progresses at pace. This is the only species of aquaculture significance for which commercial microarrays are available. The use of heterologous microarrays requires rigorous validations although excellent correlations have been observed from human gene chips hybridized with distantly-related mammalian species (bovine, porcine and canine), the Arabidopsis chip for cross-species studies and a cichlid chip for research with various cichlid species, a salmonid, poeciliid and cyprinid fish (Becher et al., 2004; Ji et al., 2004; Renn et al., 2004).

Other than species selection, decisions must be made regarding the choice of strain, ploidy status, age, sex and environmental and dietary conditioning. The acceptance of a tissue type, or part thereof, to be examined depends upon the postulated effect of a specific nutrient, dietary component or other treatment upon the target animal; the latter of which may be affected ultimately by dose and route of administration (McLean and Craig, 2003).

Essential foci for GEP in aquaculture will be animal health and welfare- and nutrition-related and for selective breeding. This is because on a global basis disease causes billions of dollars in losses and consumers demand meat products untainted by chemicals and antibiotics. Therefore, development of feed-based immunoprotectants, especially those that are naturally occurring, represents a rational approach to combating disease and a substantial reduction of reliance upon chemicals. On the other hand, feed represents the single most important operational variable in aquaculture, often exceeding 50% of a facility's annual budget. Development of methods that reduce feed costs, optimize nutrient utilisation or provide a means of adding value to final products (e.g., functionality, organic products) would be of high economic value to the industry. Another area of concern for the sustainable development of mariculture is the supply of high quality juveniles. A more comprehensive understanding of developmental nutritional physiology, nutritional requirements and the potential beneficial effects of feed additives upon the survival and weaning of larval fishes would have immense implications for aquaculture.

Biomarkers

Because microarray technologies provide exquisite empirical screening tools, clearly the conformist dogma of hypothesis-driven research no longer remains valid. Studies can be undertaken without a starting hypothesis and, following acquisition of GEPs, one may convert back to more banal hypothesis-based research. Once raw hybridization intensities have been obtained from microarray studies, and following background correction and normalization, a multitude of genes may be identified as being overexpressed.

Some of these may represent potential biomarkers of a specific process or treatment effect – that is, they provide characteristic signatures associated with a particular physiological status. In fact, a major issue requiring resolution is the current lack of validated biomarkers that are specific, sensitive, predictable and quantifiable. Once potential biomarkers are revealed by microarray analyses, their applicability and value must be validated using other methods such as quantitative real-time reverse-transcriptasepolymerase chain reaction (qRT-PCR), Northern or Western blotting or via quantification of protein levels by enzyme-linked immunosorbent (ELISA) or similar assays.

How can Aquaculture Benefit from The 'Omics' Revolution?

Already, several studies have employed the microarray-based technology to examine various aspects of fish biology. Many of these have used the zebrafish model (Crollius and Weissenback, 2005), due to the commercial availability of a microarray for this species. Although not of aquaculture importance, work with zebrafish nevertheless provides crucial information, since GEPs following various manipulations have established biomarkers that may be appropriate for research with aquaculturally important species (Table 2).

These include biomarkers for disease processes such as mycobacteriosis (Meijer et al., 2005), sex determination and reproductive function (Alberti et al., 2005; Wen et al., 2005), larval development (Ton et al., 2002; Lo et al., 2003; Linney et al., 2004), hypoxia (Ton et al., 2003) and temperature stress (Malek et al., 2004).

Microarray technology has also been applied in salmon studies (Rise et al., 2004a; 2004b; von Schalburg et al., 2005). In-house microarray technology has been developed and used to examine the early development of gilthead sea bream (Sarropoulou et al., 2005) and to evaluate the impact of a recombinant VHS vaccine upon immunity in the olive flounder (Byon et al., 2005).

Studies with Atlantic salmon and olive flounder incorporated RT-PCR verification of several maker genes (Table 2). However, it is equally noteworthy that signature biomarkers, for example, for different disease processes vary, suggesting that universal signatures for disease may be difficult to acquire. Nevertheless, diagnostic microarrays that examine a suite of genes that are intimately engaged in disease processes will undoubtedly appear in the future. Such microarrays would provide the means to assess the overall health of farmed stock and perhaps provide an advanced-warning system based upon early-stage homeostatic dysfunctions detected at the molecular level.

Electron microscopic examinations have established that the latter provide benefit in terms of optimizing intestinal integrity (Dimitroglou et al., 2005). The same might also have been assessed using marker genes associated with gut cell cycling, including claudin 6, keratin 4, myosin heavy chain 11 and others.

Mannan oligosaccharides also modulate the immune response of animals (Bland et al., 2004) and, in trout and carp, can positively affect growth, survival, and feed efficiency, while enhancing the activity of lysozyme and complement proteins (Staykov et al., 2005). Innate immunity-associated genes, including those for complement and lysozyme, have already been employed as biomarkers in mammalian and fish-based microarray research (Machado et al., 2005; Table 2). These clearly represent excellent candidates as immune markers in the aquaculture arena.

Other dietary ingredients that might be used during aquaculture, both to the benefit of the cultured organism and consumer, include various micronutrients. For example, the health benefits of selenium (Se), especially with respect to cancer risk, and its ability to withstand viral insults, are already established (Hill and Burk, 2001) and the potential exists to increase fish fillet selenium levels as a method for enhancing dietary selenium intake in humans. Hyperaccumulation of selenium in fish fillets can be attained through dietary manipulation (Cotter et al., 2005), but discrete biomarkers are still required to monitor selenium absorption and effects at different concentrations both in the host and target organism. Microarray technology has demonstrated Se-driven up- and downregulation of a broad variety of genes, including those related to detoxification, Se-binding proteins, some apoptotic genes, and genes engaged in cell proliferation (El-Bayoumy and Sinha, 2005).

Other potential selenium biomarkers have already been established for fish (Thisse et al., 2003) using microarrays, and include glutathione peroxidase, thioredoxin reductase and the cell cycle regulator p53. These and other biomarkers can be used to assess selenium status rapidly as well as to further unravel the molecular mechanisms of action of this micronutrient. Similarly, GEP can identify useful marker genes to assess micronutrient status and requirements, as well as identifying hitherto unknown beneficial (i.e., health) and negative effects of dietary nutrient concentrations.

Since approximately 40 micronutrients are required in the diets of growing animals, the savings in time, cost and animal use, and thence equipment and facilities for traditional research that are offered by transcriptomics cannot be undervalued. The same is true for macronutrient research.

Fewer than 10% of all aquacultured organisms have experienced any form of selective breeding (Gjedrem, 2005) and in general, selection programs have centred attention upon growth rates, feed efficiencies, and resistance to disease.

In contrast, selection programs for flesh quality are rare. This likely reflects the inherent difficulties and costs associated with sampling potential breeders and may be due to the fact that heritability for some important flesh quality attributes are moderate to low (0.1-0.2; Quinton et al., 2005), which likely results from polygenic effects: that is, quality traits are influenced by a large number of genes that each impart minor effects.

An alternative to traditional breeding programs is to elucidate the biological processes that control flesh quality using a markerbased approach. Improvements to flesh quality include not only manipulating traits such as tenderness or colouration, but also increasing product uniformity. This is not a trivial matter since fish flesh is processed and employed in a wide variety of ways.

Furthermore, it is not beyond the realm of possibility that in the future processors and retailers will demand the presence/up-regulation or absence/down-regulation of a series of health- and quality-related gene sets. Already the use of biomarkers to authenticate seafood product origin is employed (Piñeiro et al., 2003). Since flesh quality traits are influenced by genetic potential, environment, and feed, then a nutrigenomic approach clearly represents the most cost-effective method for enhancing product quality through rapid screening and identification of genes that influence desirable characteristics.

Apolipoproteins and their receptors play a central role in lipid metabolism, mediate the absorption of lipoproteins, regulate the cholesterol content of peripheral tissues through the reverse cholesterol transport pathway, and are intimately involved in maintenance of the nervous system and in cellular signaling. From this study, a number of genes were identified as holding potential as biomarkers, including the baculoviral IAP repeat-containing 5B (e.g., birc5b; survivin 2), which was strongly expressed in tilapia fed on high lipid diets. Birc5b is strongly associated with apoptosis.

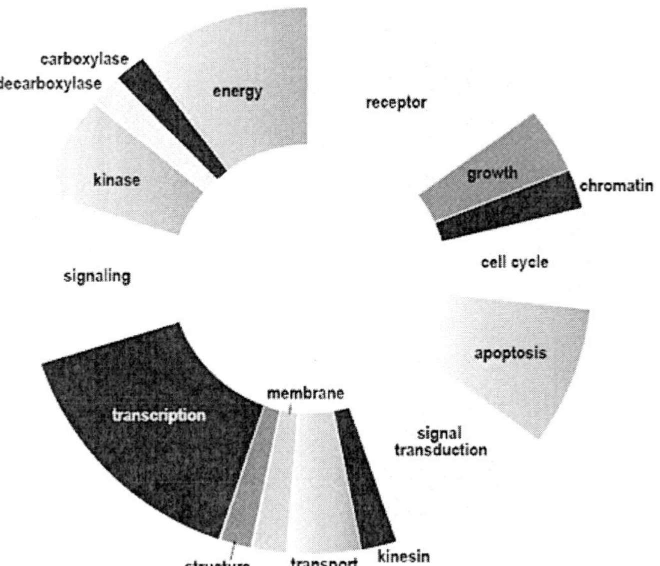

Figure: *A study designed to compare the effect of low and high energy (lipid) diets upon tilapia performance revealed that both diets influenced gene expression within the gut. Here, functional gene categories affected by the high lipid diet are illustrated for the anterior intestine. Noteworthy was that most upregulated genes were of unknown function. Approximately 33% of genes with known or inferred functions expressed lipid-induced up-regulation. In particular, apolipoprotein receptor E gene expression increased.*

Nutrigenomic approaches clearly herald a new era for the aquafeed ingredient industry. This area provides enabling technologies that permit development, refinement and modification of bioactive and natural compounds that more accurately target specific diseases or provide other health benefits both to cultured animal and consumer.

Nevertheless, it should be reiterated that great care must be taken in the analysis of generated data sets as well as in the use of biomarkers.

Since biological processes are rarely monogenetic, the predictive value of specific biomarkers may be undermined by the action of many hundreds or thousands of different genes, environmental and/or nutritional history. Even so, it is clear that the power of this technology will, when fully engaged, affect all aspects of aquaculture production, its sustainability and economic viability. As well, development of diagnostic arrays will provide exquisite techniques for establishing point of origin of farmed product as well as safety for the consumer.

It is rare indeed for aquaculture to be on a level playing field with the more established and better-funded animal production sciences. The use of this aquanomic technology catapults the discipline to the forefront of animal nutritional research.

Mendel's Laws

Hybridological analysis is one of the main methods applied in modern genetics. The honour of having created this branch of study belongs to Gregor Mendel, who published the results of his investigations in the field of hybrids of different varieties of peas (Pisum sativum) in 1865 (Mendel, 1866). Despite its brevity, this work played an outstanding part in the development of genetics; however, it outpaced the level of development of contemporary biology and remained scarcely noticed by the leading scientists.

Though his contemporaries did not appreciate the importance of the regularities established by Mendel, 35 years later, in 1900, the dialectics of these factors unquestionably brought the science to the stage of re-establishment of the regularities discovered by him.

Monohybrid Crossing

When planning his experiments, Mendel was especially careful in the selection of the parental pairs for the hybrids. Plants for the experiment had to meet the following requirements:

i. Invariably differing characters;
ii. Their hybrids should be immune to the influence of foreign pollen, or capable of being protected; and
iii. Hybrids and their descendants should not suffer from marked violation of fertility.

Unlike earlier investigators who studied simultaneously many hereditary characters, Mendel limited his studies to the behaviour of only one character at a time. This simplified considerably the problem of analyzing the behaviour of several characters in generations of hybrids. Thus, when tracing the behaviour of a character such as colour in flowers, he, for the time being, did not take into account the shape of seed, height of the plant, etc.

The method Mendel used in his work consisted of crossing two plants differing in one pair of contrasting characters. Later, such crossing, in which parental forms differ only in one pair of contrasting or alternative characters, became known as monohybrid. Then the seeds so obtained were planted, and the appearance of the first hybrid generation F_1[1] was analyzed. Then Mendel crossed two hybrid plants (or permitted them to self-pollinate) and raised as many as he could of the second generation of descendants - F_2. These descendants showed higher or lower variability in respect to the character under study, and Mendel classified them accordingly, counting the number of plants which possessed each of the two contrasting characters.

Later, Mendel crossed not only the plants differing in one, but in two or more characters (dihybrid and polyhybrid crossing). He traced the distribution of these characters in the first and second hybrid generation. Based on the practical information obtained, Mendel made certain generalizations, which were later confirmed by numerous experiments with other plants and animals. Today these generalizations have been fully confirmed and are known in genetics as Mendel's Laws of Heredity. The parental generation is denoted with "P" (parental), the first generation after crossing - "F_1" (filial), the second - "F_2" etc.

Law of Dominance, or Uniformity of The First Hybrid Generation (F_1)

One of the first factors noticed by Mendel (although the phenomenon had been known even earlier) was that out of two alternative characters of the first generation hybrids (F_1) only one usually developed. The second feature seemed to disappear or be suppressed. Thus, studying smooth and rugose forms of beans, Mendel found out that all beans of F_1 plant were smooth.

Mendel called the characters, such as revealed in beans with a smooth form, dominant; and those with rugose form, recessive. For all seven characters studied by Mendel, the domination of one over another was complete for each pair of characters, and this fact facilitated considerably his further work.

In subsequent investigations carried out by other scientists, many characteristics were found which showed complete and incomplete dominance. However, the importance of the phenomenon discovered by Mendel consisted of the fact that the appearance of a specimen could no longer be considered as an exact indicator of its hereditary constitution, because domination resulted in a hybrid or cross having the appearance of a pure-bred specimen. This is of importance in the application of the principle of dominance in cultivation of plants and in stock breeding. Later this principle of domination found its expression in such terms as phenotype (appearance) and genotype (hereditary constitution).

Law of Segregation and Independent Distribution of Characters

Mendel continued his experiments further, beyond F_1, and produced the second generation, F_2, by means of crossing F_1 plants (or by self-pollination). By subjecting all experiments to thorough qualitative analysis of the developing features, Mendel established that while in F_1 only one of the two contrasting characters was observed, in F_2 the characters appeared together in definite numerical ratio of 3 dominant to 1 recessive.

Thus, 253 smooth beans of plant F_1 gave 7 324 beans in F_2, out of which 5 474 were smooth and 1 850 rugose, i.e., a ratio very close to 3:1 (2.96:1).

In another series of experiments, where seed colouration was taken into account -yellow (dominant) and green (recessive) - out of 8 023 beans produced 6 022 appeared to be yellow and 2 001 green, a ratio again close to 3:1 (3.01:1). Further, out of 1 181 plants, 882 were ordinary convex beans and 299 were beans with cervix (a ratio 2.95:1). Mendel analyzed the behaviour of characters in the F_3 hybrid generation and discovered that if a recessive character remained constant, then seeds having dominant characters could be separated into two groups. One remained constant and imparted to the descendants only this feature, and another group behaved like F_2 hybrids, i.e., split into dominating and recessive forms in the ratio 3:1. This was a very important conclusion concerning heterogeneity of the F_2 dominating group.

Mendel also discovered that the ratio of constant segregating seed was 1:2. Thus, Mendel arrived at the conclusion that the 3:1 ratio pertaining to the distribution of dominant and recessive characters divides, therefore, for all experiments to a ratio of 2:1:1 ..... hybrids with two-character seed form differing plants, half of which remain constant and produce equal dominant and recessive characters.

The behaviour of each pair of differing characters in hybrid combination does not depend on other differences in both original plants (Mendel, 1866). This conclusion means that the distribution of hereditary factors, introduced by male and female organisms during the formation of gametes, occurs sometimes in accordance with the law of probability; i.e., the distribution by the gametes of one allelic pair is not connected (does not correlate) with the distribution of the other.

Law of Purity of Gametes

Mendel suggested an explanation for the phenomenon of segregation, and basically this explanation was correct, so that further investigations added only details, and the basic explanation itself remained unchanged. First of all, Mendel suggested that the elementary cells possessed properties or factors, the combination of which gave rise to one or another character. It must be stipulated that Mendel himself did not draw a clear line between a hereditary factor (according to modern terminology - gene) and a character. However, the logic of his experiment did distinguish between these two notions; the conclusion Mendel reached, on the basis of his own data, showed a clear difference between these two. Mendel reached the conclusion that hybrids form elementary pollen cells, which by their properties and to an equal extent, correspond to all the constant forms developed from the combinations of characteristics resulting from pollination (Mendel, 1866).

In other words, Mendel reached the conclusion that during the formation of gametes by a hybrid specimen, bearing, for example, factors of smooth and rugate seed, these gametes are not similar, and they give rise to two different types, half bearing only the smooth seed factor, and the other bearing the rugate seed factor.

This brings about two very important stipulations. First, each gamete is pure (not hybrid), i.e., contains one member of each pair of factors. Second, during gamete formation, the hereditary substance decreases to half; thus each gamete contains only half of the factors which are present in the cells of the parent plant. Consequently, the main peculiarity of the mechanism of splitting relates to the two factors, one of which is brought by the gametes of one parent, and the other by the gametes of the second parent; these combine, and during the whole life time of a generation, exist together in the cells of a hybrid specimen, without crossing and losing their identity. When in such a hybrid its own sex cells are formed, these two factors again segregate or split, and each gamete contains only one of these factors and is pure (that is free from another alternative factor).

8

Selection for Qualitative Phenotypes

Qualitative phenotypes are usually less important than quantitative phenotypes for food fish farmers, but an understanding of how they can be controlled and exploited by selective breeding programmes is important for two reasons: First, and most important, is the fact that qualitative phenotypes can affect the value of the crop or the cost of production. If a farmer can produce a more attractive product, consumers are often willing to pay more for it, which increases his profits. On the other hand, some qualitative phenotypes, such as deformities, can decrease the value of the population, either by increasing the cost of production or by producing fish so unattractive that consumers will not buy them. In both cases, farmers can use selection to improve their populations. Secondly, an understanding of how selective breeding programmes can be used to fix (frequency = 100%) the desired qualitative phenotypes and eliminate (frequency = 0%) the undesired ones and thus produce true-breeding populations may make it easier to understand how selection can be used to improve quantitative phenotypes. Those who already have a good grasp of Mendelian genetics and understand how breeding programmes can be used to fix or to eliminate qualitative phenotypes, or those who only wish to work with quantitative phenotypes.

Factors that must be Considered before Conducting Selective Breeding Programmes

Before conducting a selective breeding programme, a farmer must realise that he can use selection to accomplish only those goals which are achievable biologically. Selection is a breeding programme that exploits heritable phenotypic variance; consequently, the two prerequisites for a successful selective breeding programme are: one, the phenotype that the farmer wants to change must exhibit variance; two, the variance must be heritable. Breeders are able to accomplish only what the fish and the fish's genes allow them to accomplish. Breeders cannot custom-design fish by removing undesirable phenotypes, such as sharp pectoral spines, unless individuals without these spines are discovered, and even if such individuals are discovered, the phenotype must be heritable or selection will be an expensive, futile, and frustrating waste of effort.

If the phenotypes in question are heritable, the genetics should be known before a breeding programme is conducted. This information will enable the breeder to use the most appropriate breeding programme. Finally, the farmer must know the relative costs of the traits (production costs) and what they are worth (market value) before he initiates a selective breeding programme.

Qualitative phenotypes can be described as "either/or" phenotypes. A fish either has one phenotype or it has another; in other words, the fish in a population fall into discrete, non-overlapping phenotypic categories. Because selection exploits phenotypic variance, at least two phenotypes must exist in the population, and one must be the phenotype that the farmer wants. If there is only one phenotype, there is no phenotypic variance, which means that selection cannot be used to alter the phenotype. For example, if all fish in a population are black, there is no phenotypic variance for body colour; consequently, selection cannot be used to change body colour.

Not only must variance exist, but the variance that is observed must be heritable, or selection will not be able to alter phenotypic frequencies. The distinction between heritable and non-heritable variance is important, because many qualitative phenotypes that are observed at fish farms and at fish hatcheries are non-heritable deformities. Some deformities have a genetic basis, but most are caused by nutritional deficiencies, environmental disturbances, toxins, disease, or injury or they are developmental mistakes. The deformities that are produced by non-genetic factors cannot be eliminated by selection. The only way to eliminate a non-heritable deformity is to discover the environmental factor that causes it and to eliminate the factor or modify the culture environment. It is often difficult, if not impossible, to identify the cause of a non-heritable deformity, so unless the frequency approaches 0.5% it probably should be ignored.

Before selection is used to fix or to eliminate qualitative phenotypes and produce true-breeding populations, the way the phenotypes are inherited should be determined. If a previous experiment has already determined the mode of inheritance, that information can be used to design the breeding programme. If the mode of inheritance is known, the correct breeding programme can be chosen, and this will enable the farmer to achieve his goal quickly and efficiently. If the mode of inheritance is not known prior to the breeding programme, the farmer may choose the wrong programme, which means he will not accomplish his goal and will waste effort, facilities, and money.

The research programmes that are needed to determine the modes of inheritance for qualitative phenotypes are not that complicated. In general, the inheritance is deciphered by conducting paired matings across two generations and by determining the phenotypic ratios produced by these matings. Even though this type of research is not that complicated, it requires effort, facilities, and the ability to analyse the results statistically. Such research should be left to scientists at universities or governmental research stations.

Before conducting a selective breeding programme, a farmer should conduct an assessment to determine if a selective breeding programme is needed. The assessment can be conducted by a farmer (such studies are not cheap, and they require a high level of sophistication), or they can be conducted by fisheries cooperatives or a local fish farming research station. The farmer needs to determine if his customers want fish with a different body colour, or if they will pay extra for a prettier fish. His assessment will also tell him if the proposed breeding programme could open new markets. A farmer should only conduct a selective breeding programme when it is necessary or when it will increase his profits. Changing body colour or other phenotypes might produce a prettier fish, but if the assessment reveals that consumers do not want a different body colour, the breeding programme should not be conducted.

Finally, the farmer must know the costs of producing the alternate phenotypes before conducting a selective breeding programme. The values of some phenotypes are so obvious that no formal scientific study is needed. If a farmer wants to use selection to eliminate a heritable deformity which lowers growth rate or viability, formal studies to determine the relative values of the phenotypes are unnecessary, because he already knows that the deformity costs him money. On the other hand, the relative production costs of some phenotypes are more difficult to determine and require evaluation in a scientific experiment. The relative production costs are determined by assessing the effects that the phenotypes have on growth rate, survival, egg production, etc. These secondary effects are called "pleiotropic effects," and in agriculture and aquaculture they can be more important than the phenotype itself—especially if growth, survival, or fecundity are affected. Almost all mutant body colours that have been investigated in fish have exhibited negative pleiotropic effects, in that they adversely affected growth rate or survival.

If there are no negative pleiotropic effects or if the extra costs of production (slower growth rate and/or lower viability) are offset by the increased market value of the crop (extra money consumers are willing to pay), selection is worthwhile. On the other hand, if the increased costs of production exceed the extra market value of the crop, it would be foolish to conduct a selective breeding programme because the farmer will earn less money.

Selective Breeding Programmes to Produce True-Breeding Populations

When working with qualitative phenotypes, the goal of all selective breeding programmes is to produce a true-breeding population. A true-breeding population is one that contains only one allele for the locus in question because it was fixed; selection eliminated the other allele. This means that every gamete produced by the select brood fish will contain only the desired allele, which means that every offspring produced by the select brood fish will have the desired phenotype. The only way the undesired phenotype can be produced after selection has been used to create a true-breeding population is by the accidental (or intentional) stocking of fish from another population or by mutation.

The undesired phenotype and the allele that produces it are eliminated by a process called "culling." Culled fish are not allowed to reproduce. They do not have to be killed; they can be grown and then sold as food, but they must not be allowed to spawn.

There are three basic rules that govern the selection of qualitative phenotypes:

1. If the desired phenotype is controlled only by a homozygous genotype, a single act of selection will create a true-breeding population.

2. If the desired phenotype is controlled by two or more genotypes, selection cannot create a true-breeding population. Progeny testing must be used to fix the desired allele and produce a true-breeding population.
3. If the desired phenotype is controlled by the heterozygous genotype, no breeding programme can create a true-breeding population. A population in which every fish has the desired phenotype can be produced by crossing the two homozygous phenotypes, but this procedure must be repeated every breeding season.

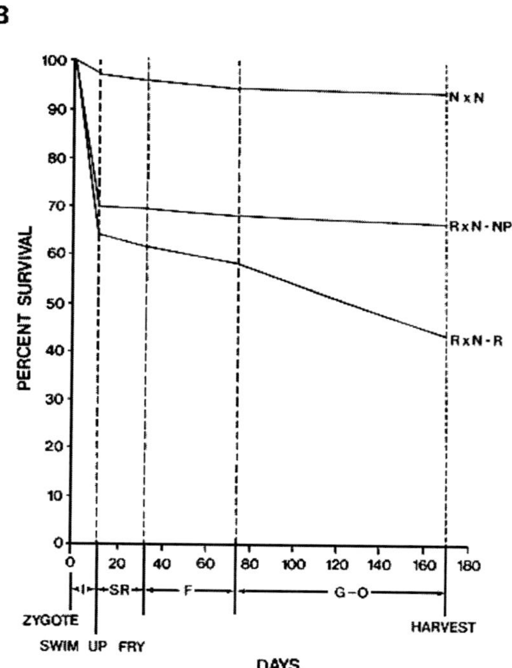

Figure: *Effect of red body colour on viability in Nile tilapia. This is an example of a negative pleiotropic effect. Viability or red fish (R × N-R) is compared to that of their normally pigmented sibs (R × N-NP) and a control population which had no red fish (N × N). Viabilities during artificial incubation of eggs (I), sex reversal (SR), fingerling production (F), and grow-out (G-O) are shown. At harvest, fish averaged 250 g.*

Source: After El Gamal, A.R.A.L. 1987. Reproductive performance, sex ratios, gonadal development, cold tolerance, viability and growth of red and normally pigmented hybrids of Tilapia aurea and T. nilotica. Doctoral dissertation, Auburn University, Alabama, USA.

Selection for Homozygous Phenotypes

When this type of selection is used, it is usually done because a farmer wants to fix the recessive phenotype and eliminate the dominant phenotype, and the phenotypes are produced by an autosomal gene that exhibits complete dominance. The term "recessive" does not imply inferiority; similarly, the term "dominant"does not imply superiority. The terms simply refer to the way the alleles express themselves and to the phenotypes that are produced by those alleles. Many recessive phenotypes have been shown to be quite valuable in various farmed plants and animals, and breeding programmes have been conducted to fix them. The terms "inferior" and "superior" should be applied to phenotypes only after biological and economic evaluations have been conducted.

If the desired trait is the recessive phenotype that is produced by a gene that exhibits complete dominance, the selective breeding programme that is needed to fix the trait and to produce a true-breeding population is simple, and this goal can be achieved by a single act of selection. To accomplish this, fish with the recessive phenotype will be saved, and fish with the dominant phenotype will be culled.

Genetically, fish with the undesired dominant phenotype are either homozygous dominant or heterozygous. Because every fish that has at least one dominant allele expresses the undesired dominant phenotype, a simple one-step culling of all fish with the dominant phenotype will eliminate all copies of the dominant allele. Conversely, every fish that expresses the desired recessive phenotype is homozygous recessive, which means that when these fish are selected, none will possess a copy of the undesired dominant allele. Thus, the recessive allele will be the only allele that exists in the select population, which means that it will breed true.

For example, say a tilapia farmer has both normally pigmented and pink Nile tilapia, and decides to grow only pink tilapia. He can create a true-breeding pink population by a simple one-step selective breeding programme during which he will cull all normally pigmented fish. Pink body colour is the recessive phenotype, and it is produced by the recessive b allele (bb genotype); normal body colour is the dominant phenotype, and it is produced by the dominant B allele (BB and Bb genotypes).

The culling of all normally pigmented tilapia will remove all B alleles from the population. The only fish that remain will be pink and, since they are homozygous recessive bb, the b allele will be the only allele that will remain in the population of select brood fish. Consequently, a single act of selection will create a true-breeding pink population.

This type of selective breeding programme can also be used to fix either the recessive or the dominant phenotypes when the mode of gene action is incomplete dominance, and it can also be used to fix either homozygous phenotype when the mode of gene action is additive. In both cases, there are two homozygous genotypes that can be identified and isolated. By culling the undesired phenotypes, true-breeding populations can be created by a single act of selection.

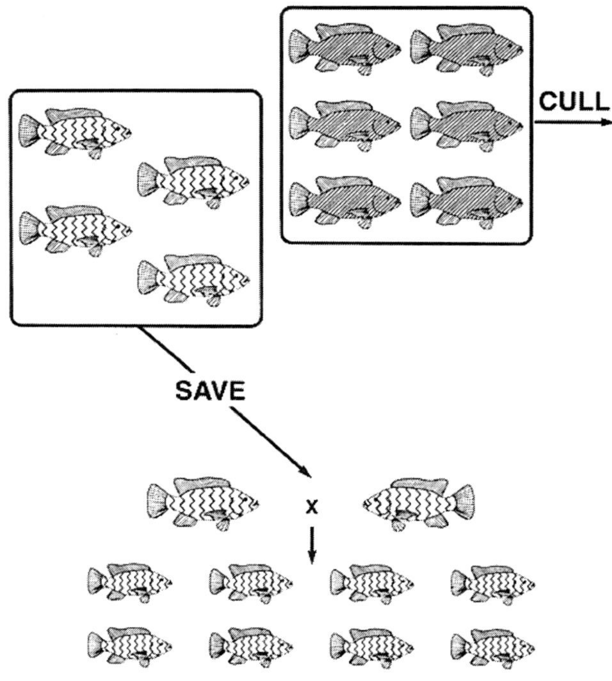

Figure: *Selective breeding programme needed to produce a true-breeding population of pink Nile tilapia. If all normally pigmented (the dominant phenotype) fish are culled, all copies of the dominant B allele will be eliminated. The only fish that remain are pink (the recessive phenotype), and since they are homozygous recessive bb, the select population of pink brood fish will breed true and will produce only pink offspring. Any recessive phenotype can be fixed this easily.*

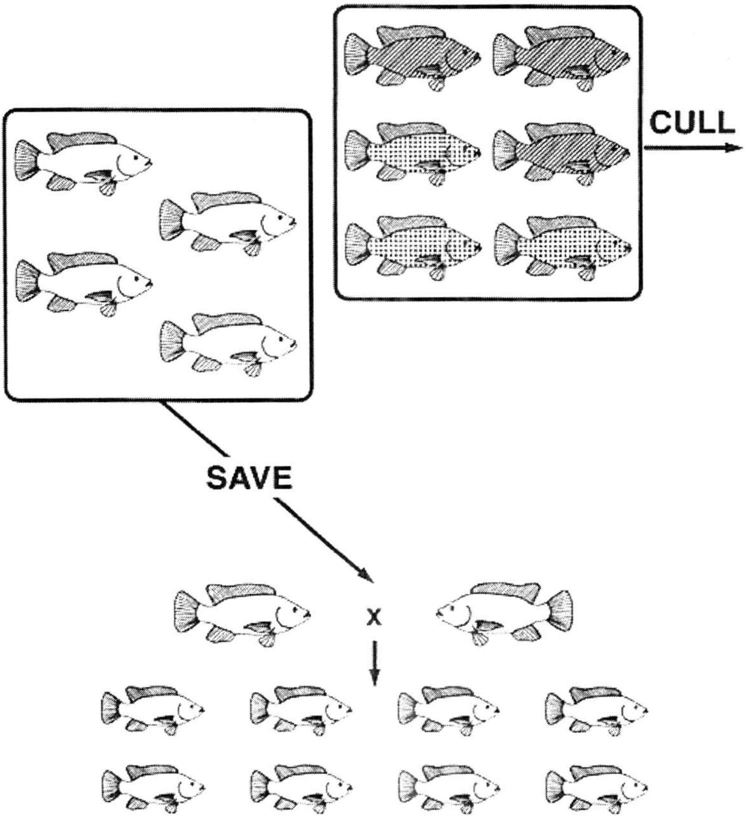

Figure: *Selective breeding programme needed to produce a true-breeding population of gold Mozambique tilapia. If all melanistic (black [GG] and bronze [Ggl]) fish are culled, all copies of the dominant G. allele will be eliminated. The only fish that remain are gold (the recessive phenotype), and since they are homozygous recessive gg, the select population of gold brood fish will breed true and will produce only gold offspring. Pictorial representations of the phenotypes are the same as those used in figure.*

For example, say a tilapia farmer has black, bronze, and gold body colours in his population of Mozambique tilapia. If the farmer wants to produce either a true-breeding population of gold or a true-breeding population of black fish, he can accomplish either goal by a single act of selection. Gold is the recessive phenotype, and it is produced by the homozygous recessive genotype (gg); black is the dominant phenotype, and it is produced by the homozygous dominant genotype (GG); bronze is the heterozygous (Gg) phenotype.

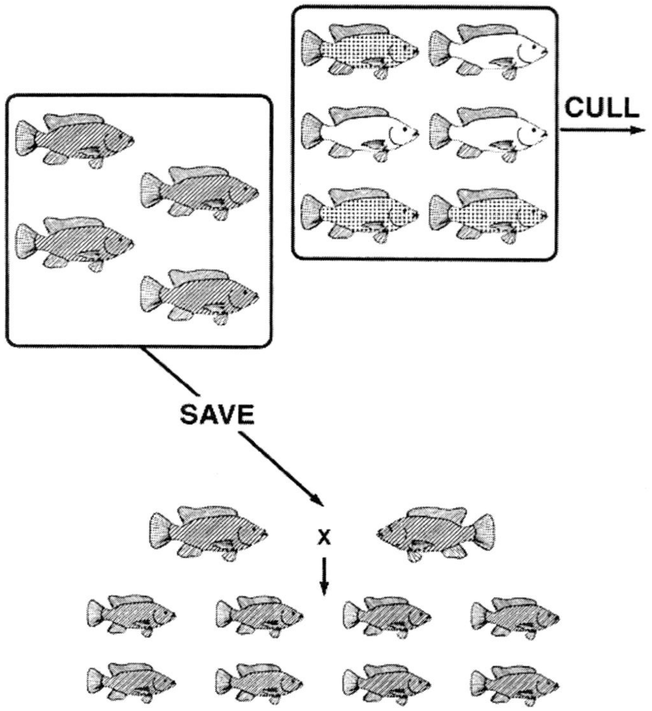

Figure: *Selective breeding programme needed to produce a true-breeding population of black Mozambique tilapia. If all bronze (the heterozygous [Gg] phenotype) and all gold (the recessive [gg] phenotype) fish are culled, all copies of the recessive g. allele will be eliminated. The only fish that remain are black (the dominant phenotype), and since they are homozygous dominant GG, the select population of black brood fish will breed true and will produce only black offspring. Pictorial representations of the phenotypes are the same as those used in Figure.*

A true-breeding population of gold tilapia, can be produced by culling all black and all bronze fish. Conversely, a true-breeding population of black tilapia can be produced by culling all gold and all bronze fish.

This same concept can be extended to phenotypes that are controlled by two or more genes. If the desired phenotype is controlled only by a homozygous genotype, a true-breeding population can be created by a single act of selection.

For example, if a carp farmer has a population of common carp with all four scale phenotypes and wants to produce a true-breeding population that has a reduced scale pattern, all he has to do is save the fish with the mirror phenotype and cull all other fish. Three of the scale phenotypes have a reduced number of scales, but only mirror is produced by a homozygous genotype (ss, nn), so it is the only one that is capable of breeding true. Both the leather (ss, Nn) and line (SS, Nn and Ss, Nn) phenotypes are undesired, because they are heterozygous at the N locus, which means that when two leather and/or line fish mate, 25% of their offspring will die (those that are homozygous NN. regardless of the genotype at the S locus). Leather and line phenotypes are also undesired, because fish with these phenotypes have several negative pleiotropic effects, among them lowered growth rates and viabilities.

If the farmer culls all leather, line, and scaled individuals, all S and N alleles will be eliminated. The only fish that will remain in the select population will be those with the mirror phenotype, and since they are homozygous recessive (ss, nn) the s and n alleles will be the only ones that remain in the population of select brood fish. Consequently, a single act of selection will create a true-breeding population of mirror common carp.

Phenotypes Controlled by More than One Genotype

Selection cannot fix a phenotype that is controlled by more than one genotype. This means selection cannot fix a dominant phenotype and create a true-breeding population if it is produced by a gene with complete dominance.

Conversely, this also means that selection cannot eliminate a recessive phenotype (and the recessive allele) if it is produced by a gene with complete dominance. Despite the fact that this type of selective breeding programme will not work, it is routinely used in a misguided effort to eliminate undesired phenotypes that are produced or that are assumed to be produced by recessive alleles.

This type of breeding programme will not achieve its goal for either of two reasons. The first is the fact that most undesired phenotypes are abnormalities that are created by non-genetic factors (environmental disturbances or developmental mistakes). Many assume that all abnormal phenotypes are mutant phenotypes which are produced by recessive alleles. Some are, but most are not. If the undesired phenotype is produced by a non-genetic factor, selection will not work, because no selective breeding programme can fix or eliminate non-heritable phenotypes.

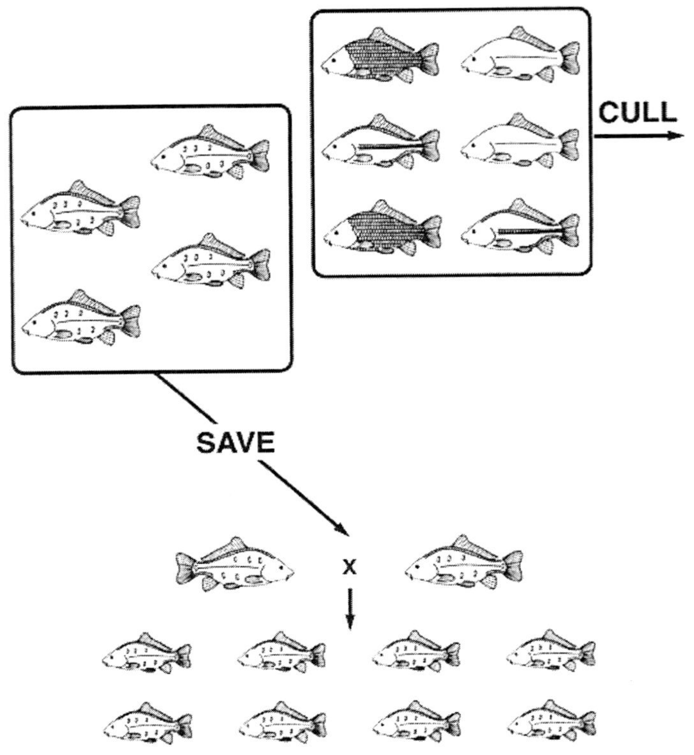

Figure: *Selective breeding programme needed to produce a true-breeding population of mirror common carp. If all scaled, line, and leather fish are culled (phenotypes that are produced by genotypes with at least one dominant allele), all copies of the S and N alleles will be eliminated. The only fish that remain are those with the mirror (the recessive) phenotype, and since they are homozygous recessive ss, nn the select population of mirror brood fish will breed true and will produce offspring with only the mirror phenotype. Even if a set of phenotypes are produced by two or more genes, if the desired phenotype is the recessive phenotype (all loci responsible for the production of the phenotype are homozygous recessive), selection can produce a true-breeding population this easily.*

Even when a deformity is produced by a recessive allele, if it is an autosomal gene that exhibits complete dominance, selection will not create a deformity-free population. The reason why it will fail is quite simple: The dominant (normal) phenotype is produced by either of two genotypes, and it is impossible to differentiate normal fish that are homozygous dominant from those that are heterozygous.

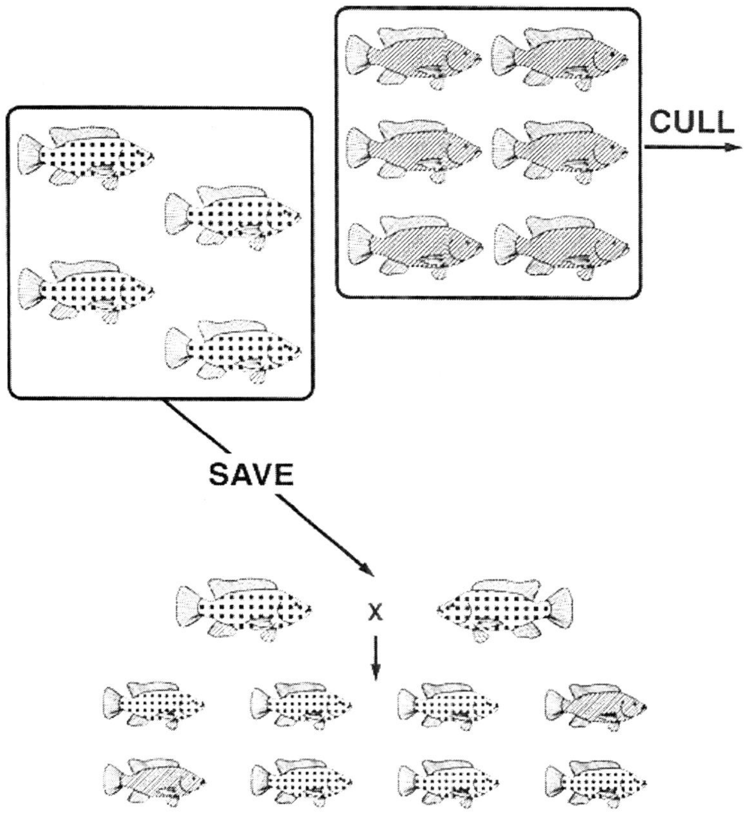

Figure: *Selective breeding programme used in an unsuccessful attempt to produce a true-breeding population of red Nile tilapia. Red fish are represented by fish covered with squares, while normally pigmented fish are those covered with stripes. If all normally pigmented fish (the recessive phenotype) are culled, the select population will be composed of only red fish (the dominant phenotype). Since red body colour is produced by a dominant allele that exhibits complete dominance, red is produced by both the homozygous dominant (RR) and by the heterozygous (Rr) genotypes. Every heterozygous red brood fish (Rr) that is saved will carry a copy of the r allele to the select population; consequently, selection cannot eliminate it. When select red brood fish are spawned, the mating of two heterozygotes (Rr) will produce some normally pigmented offspring (the mating that is depicted). Recessive phenotypes cannot be eliminated by culling them; they will reappear when the select brood fish are spawned. Consequently, if the desired phenotype is the dominant phenotype and it can be produced by both the homozygous dominant and by the heterozygous genotype, selection cannot create a true-breeding population.*

Because of that, when a farmer culls the recessive (abnormal) fish, the select population will be composed of fish with one phenotype (normal) but two genotypes, and every heterozygous select normal brood fish will carry a copy of the undesired recessive allele. Because the select population contains both alleles, it cannot breed true. When two heterozygous select normal brood fish mate, they will produce offspring with the undesired abnormal (recessive) phenotype. This type of selection will reduce the frequency of an undesired recessive phenotype, but it cannot eliminate it and produce a true-breeding deformity-free population.

For example, say a tilapia farmer has a population of Nile tilapia that contains both normally pigmented and red individuals, and decides to produce only red tilapia. If he tries to produce a true-breeding population of red Nile tilapia by culling all normally pigmented fish, he will discover that he will be unable to produce a red population that breeds true. The select red brood fish will produce both red and normally pigmented offspring. This is because red is the dominant phenotype, and it is produced by the dominant R allele (RR and fir genotypes), while normal pigmentation is the recessive phenotype, and it is produced by the recessive r allele (rr genotype). Because the homozygous dominant and the heterozygous genotypes produce identical red phenotypes, some of the select red brood fish (the heterozygotes) will have a copy of the undesired r allele. When two heterozygous red brood fish mate, they will produce some normally pigmented offspring.

The same concept applies for phenotypes that are controlled by two or more genes. If the desired phenotype is produced by more than one genotype, and if at least one of the genes can be in either the homozygous or heterozygous state, selection cannot create a true-breeding population. As was the case for phenotypes produced by a single gene, some of the select brood fish will carry a copy of the undesired recessive allele(s) and, when two heterozygotes mate, they will produce offspring with the undesired phenotype(s).

For example, if a carp farmer has a population of common carp with all four scale phenotypes and decides to produce a true-breeding population of scaled common carp by selection, he will discover that he cannot create a true-breeding population. This is because the select scaled brood fish will have two genotypes—SS, nn or Ss, nn-and it is impossible to tell heterozygous scaled fish from homozygous scaled fish. The farmer can eliminate the dominant N allele by culling all fish with the line and leather phenotypes (a dominant allele can be culled by a single act of selection); those individuals with two copies of the N, allele were culled by nature and died. However, the farmer cannot eliminate the undesired s allele by culling the remaining phenotype-mirror. The select population of scaled brood fish will produce both mirror and scaled offspring when two select heterozygous (Ss, nn) scaled brood fish mate.

Progeny testing: Since selection against (culling) a recessive phenotype cannot fix a dominant allele and produce a true-breeding population (when the mode of gene action is complete dominance), another type of breeding programme must be used to achieve this goal. The only way an undesired recessive allele can be eliminated is by progeny testing. Progeny testing is a breeding programme that deciphers a parent's genotype by determining its offspring's phenotypes. Once the genotypes of the dominant parents are deciphered, selection is used to save the homozygotes and cull the heterozygotes, which will eliminate all copies of the recessive allele, fix the dominant allele, and produce a true-breeding population.

This is accomplished by pairing and mating fish with the dominant phenotype to test fish (a fish whose genotype is known). Test fish are usually fish with the recessive phenotype, because such fish are homozygous recessive (experiments are not needed to determine a recessive phenotype's genotype). Because a test fish can produce gametes with only a recessive allele, the dominant fish is the parent that determines its offspring's phenotypes. If the dominant parent is homozygous, all offspring will have the dominant phenotype; if the dominant parent is a heterozygote, half the offspring will have the dominant phenotype and half will have the recessive phenotype.

The production of a single offspring with the recessive phenotype is enough evidence to state with certainty that the dominant parent is a heterozygote and must be culled. If no offspring with the recessive phenotype is detected in a random sample of at least 20 offspring, the dominant parent can be declared to be homozygous and kept for breeding purposes. Once a sufficient number of males and females are progeny tested and declared to be homozygous dominant, all other fish (including all offspring produced in the progeny tests) are culled. The population will now breed true and produce only offspring with the dominant phenotype.

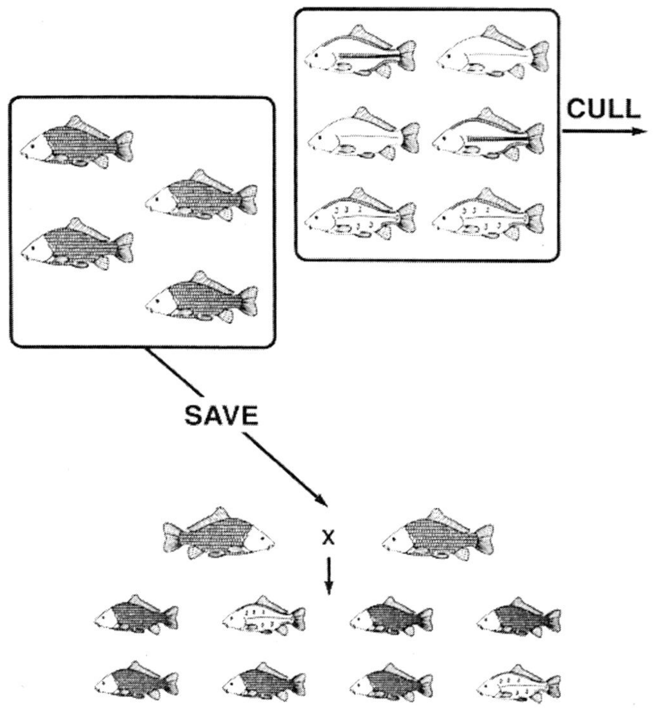

Figure: *Selective breeding programme that is used in an unsuccessful attempt to produce a true-breeding population of scaled common carp. If all fish with the line, mirror, and leather phenotypes are culled, the select population will be composed of scaled fish. This act of selection will eliminate all copies of the undesired dominant N allele, because every heterozygote (line and leather phenotypes) will be culled (the homozygotes were culled by nature, since the NN genotype is lethal). Unfortunately, this selective breeding programme will not produce a true-breeding population because you cannot eliminate the undesired recessive s allele by culling the recessive phenotype (mirror). Because the scaled phenotype is produced by a dominant allele that exhibits complete dominance, both homozygous (SS, nn) and heterozygous (Ss, nn) scaled fish will be saved. Every heterozygous select scaled brood fish will carry a copy of the recessive s allele to the select population; consequently, selection cannot eliminate it. When select scaled brood fish are spawned, the mating of two heterozygotes will produce some offspring with the mirror phenotype (the mating that is depicted). As was the case with a phenotype produced by a single gene, if the desired phenotype is produced by more than one genotype and if one of the genotypes is a heterozygous genotype, selection cannot create a true-breeding population.*

For example, a tilapia farmer that has a population of red and normally pigmented Nile tilapia can create a true-breeding red population by using progeny testing to identify and keep red fish that are homozygous (RR) and to identify and cull red fish that are heterozygous (Rr). Before initiating the progeny testing programme, the farmer will cull all normally pigmented fish except for the ones that he will use as test fish.

The farmer progeny tests the red fish by pairing a red fish with a test fish (in this case, normally pigmented tilapia are used as test fish since they are homozygous recessive rr). Red parents that are homozygotes (RR) will produce only red offspring and will be saved; red parents that are heterozygotes (Rr) will produce equal numbers of red and normally pigmented offspring and will be culled. After one generation of progeny testing, the farmer will have a select population of red brood fish that will breed true and produce only red offspring, because every select red brood fish is homozygous RR.

The same principle applies if the desired phenotype is controlled by two or more genotypes and the phenotype is controlled by more than one gene. For example, if a carp farmer has a population of common carp with all four scale patterns and decides to produce a true-breeding population of scaled common carp, he will have to use progeny testing to accomplish his goal.

Scaled fish are produced by both the SS, nn and Ss, nn genotypes. In this case, fish with the mirror scale pattern are used as the test fish because they are homozygous recessive (ss, nn). Before initiating the progeny testing programme, the farmer will cull all fish with the line and leather phenotypes and most fish with the mirror phenotypes; he will keep a few fish with the mirror phenotype and use them as test fish.

Progeny testing is done to identify and save scaled fish that are homozygous at the S locus (SS,nn) and to identify and cull scaled fish that are heterozygous at the S locus (Ss.nn). If the scaled parent is homozygous at the S locus (SS, nn), all offspring will have the scaled phenotype; if the scaled parent is a heterozygote (Ss, nn), half the offspring will have the scaled phenotype and half will have the mirror phenotype. After one generation of progeny testing, the farmer will have a select population of scaled brood fish that are all SS, nn, and it will breed true and produce only scaled progeny.

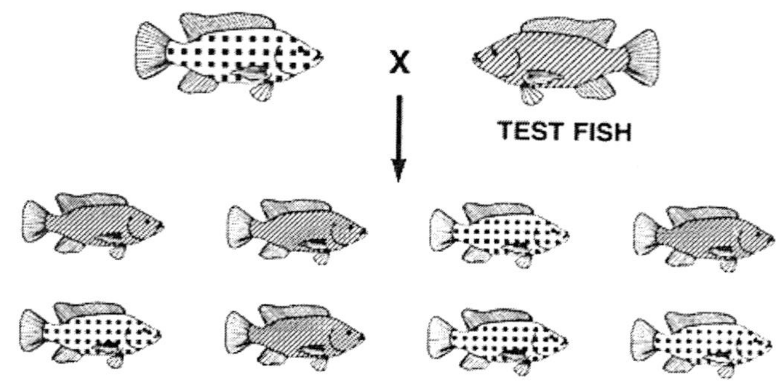

CULL RED FISH. IT IS A HETEROZYGOTE (*Rr*).

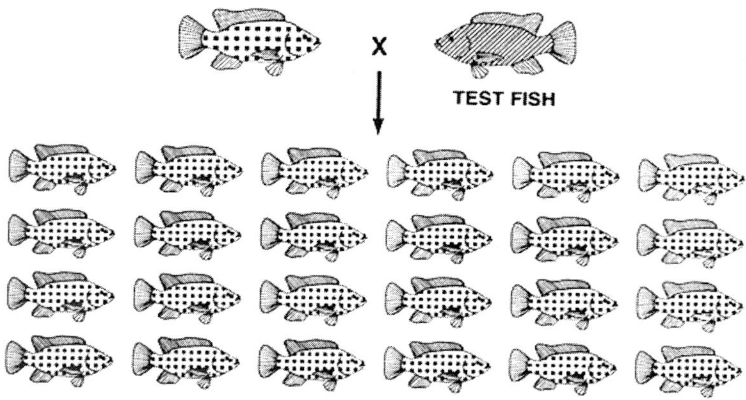

KEEP RED FISH. IT IS A HOMOZYGOTE (*RR*).

Figure: *Schematic diagram of the progeny testing programme that must be used to create a true-breeding population of red Nile tilapia. Progeny testing is used to identify and cull the heterozygous (Rr) red fish and to identify and save the homozygous (RR) red fish. A red fish's genotype is deciphered by mating it to a test fish (a normally pigmented fish [rr]), and by then determining the phenotypes of its offspring. In the top mating, the red fish is identified as a heterozygote (Rr) and is culled, because half of its offspring are normally pigmented. In the bottom mating, the red fish is considered to be a homozygote (RR) and is saved, because no normally pigmented fish was found in a random sample of a least 20 of its offspring. Progeny testing will create a population of select red brood fish that are all certified to be homozygous BE, and such a population will breed true and produce only red offspring.*

When progeny testing, if the fish can be stripped, a single test fish can be used to progeny test a number of fish with the dominant phenotype. Additionally, it is not necessary to raise thousands of individuals from each family. A random sample of 100–200 fertilized eggs should be sufficient, if survival to the early fingerling stage is good. Consequently, progeny testing does not have to be expensive; it can be done in a small hatchery in aquaria, in small plastic swimming pools, or in hapas suspended in a single pond no larger than 0.04 ha. While the families are being raised, the dominant brood fish must be kept in isolated tanks or hapas so they can be saved or culled once the progeny are examined. Several fish can be kept in a tank or hapa if they are marked by fin clips. A single generation of progeny testing will create a true-breeding population, and it will not be needed again, unless fish carrying the undesired recessive allele are acquired from another farm or the allele arises by mutation.

Selection for Phenotypes Controlled by Heterozygous Genotypes

If the heterozygous phenotype is desired, no breeding programme can be used produce a true-breeding population. This is because fish with the desired phenotype are unable to breed true. Heterozygotes produce two types of gametes and produce them in equal numbers. Consequently, when they mate, they produce offspring with three phenotypes (selection for the heterozygous phenotype can only be done for phenotypes controlled by autosomal genes with incomplete dominance or with additive gene action). A farmer who culls the two homozygous phenotypes and creates a select population containing only the heterozygous phenotype, will quickly discover that the selective breeding programme is a failure, because only half the offspring produced by the select brood fish will have the desired phenotype.

For example, if a tilapia farmer has black, bronze, and gold phenotypes in his population of Mozambique tilapia and decides to produce a true-breeding population of bronze fish, he will find that it is impossible. If he conducts a selective breeding programme, he will cull the gold (the recessive phenotype) and the black (the dominant phenotype) fish and will save the bronze ones (the heterozygotes). When he breeds the select population of bronze brood fish, they will produce offspring with all three phenotypes. If the farmer wants to produce nothing but bronze fish, he must mate gold fish with black fish. This type of mating (the mating of the two homozygotes) will produce 100% heterozygous offspring. The only requirement for this breeding programme is that the farmer must be able to accurately sex his fish. If the sex of a single fish is mis-identified, gold or black offspring will be produced. This type of mating scheme requires the use of at least two ponds (or tanks) for offspring production: one will be stocked with gold females and black males; the other will be stocked with black females and gold males.

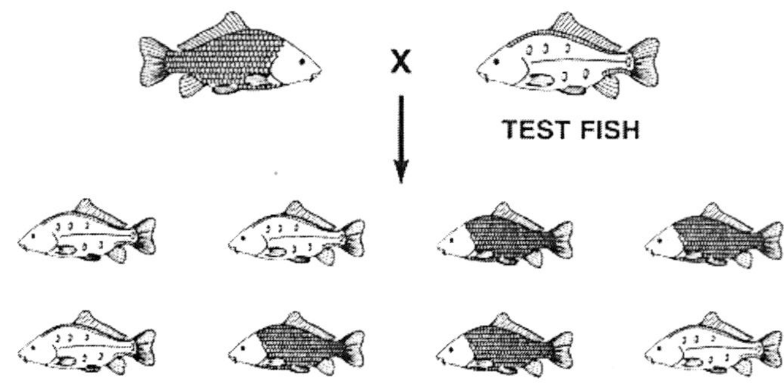

CULL SCALED FISH. IT IS A HETEROZYGOTE (*Ss*, *nn*).

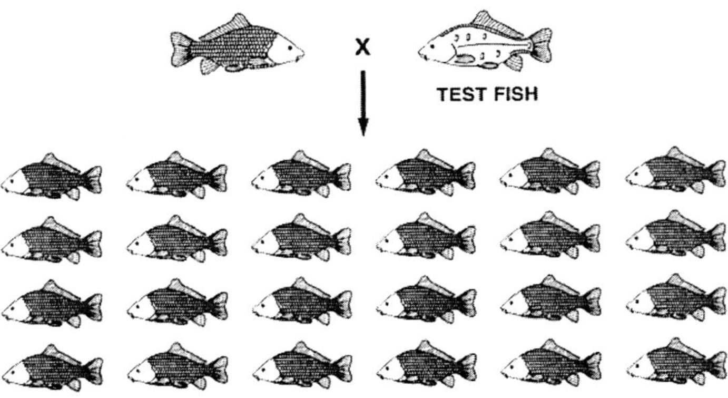

KEEP SCALED FISH. IT IS A HOMOZYGOTE (*SS*, *nn*).

Figure: *Schematic diagram of the progeny testing programme that must be used to create a true-breeding population of scaled common carp. Progeny testing is used to identify and cull the heterozygous (Ss, nn) scaled fish and to identify and save the homozygous (SS, nn) scaled fish. A scaled fish's genotype is deciphered by mating it to a test fish (a fish with the mirror phenotype [ss, nn]), and by then determining the phenotypes of its offspring. In the top mating, the scaled fish is identified as a heterozygote (Ss, nn) and is culled, because half of its progeny have the mirror phenotype. In the bottom mating, the scaled fish is considered to be a homozygote (SS, nn) and is saved, because no fish with the mirror phenotype was found in a random sample of at least 20 of its offspring. Progeny testing will create a select population of scaled brood fish that are all certified to be homozygous SS, nn, and such a population will breed true and produce only scaled offspring.*

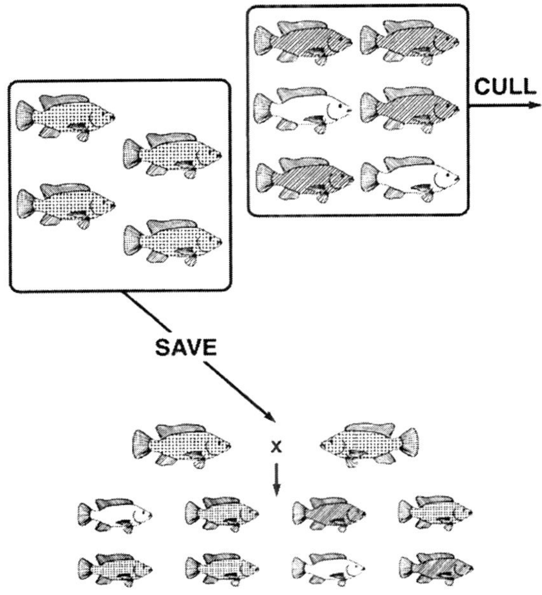

Figure: *Selective breeding programme used in an unsuccessful attempt to produce a true-breeding population of bronze Mozambique tilapia. In this case, both homozygotes-black (GG) and gold (ggl)-are culled, and the bronze (Gg) fish are saved. Because bronze fish are heterozygotes, the select population cannot breed true. Only half the progeny produced by the select brood fish will be bronze.*

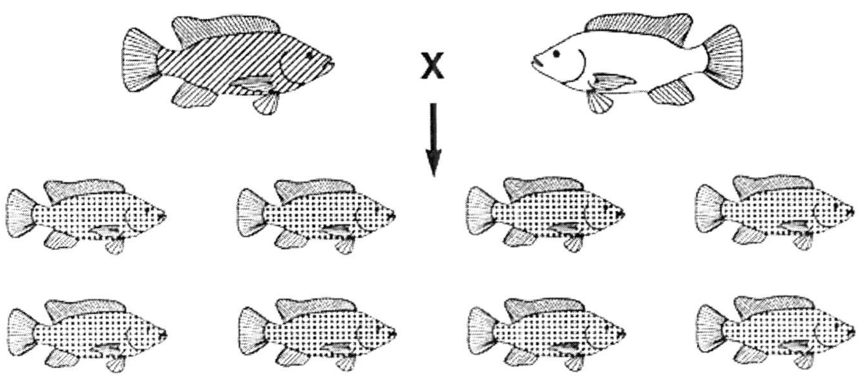

Figure: *Breeding programme needed to produce 100% bronze Mozambique tilapia. A population of bronze (Gg) fish can be produced only by mating black (GG) fish with gold (gg) fish. If a farmer wants to grow only bronze fish, he must use this mating programme every breeding season.*

Record Keeping

When compared with the volume of records that must be kept when working with quantitative phenotypes, the types and amount of records that must be taken and maintained when working with qualitative phenotypes are minimal.

One type of information that should be recorded is a description of abnormal or "mutant" phenotypes that are observed every year. Farmers' memories are not perfect, and they often feel that they saw more abnormal fish than really existed, simply because these fish are so unusual.

A census of these phenotypes will tell a farmer what types of abnormalities have been produced and will also let him know if an abnormality is becoming a problem. These data will also provide a farmer with information about other qualitative phenotypes that exist in his population.

A farmer should have a data sheet on which he records the date, the name of the phenotype (a name can be invented if none exists), and a brief description of the phenotype. In addition, the number of fish that had this phenotype and the total number of fish that were produced should be recorded. This will let the farmer know the frequencies for each abnormality. If the frequencies begin to increase over time, the farmer will know that there is a potential problem.

Table: *Example of a data sheet that can be used to record abnormal and/ or other qualitative phenotypes that are observed.*

Date: March 23, 1995 Species: Nile tilapia Number produced: 25,500			
Phenotype	Description	Number	Frequency
Tail less	No tail; no caudal peduncle	3	0.000117
Semi-operculum	Right operculum short; gills partially exposed	5	0.000196
Stumpbody	Dwarf; trunk abnormally short	2	0.000078

If the farmer is going to conduct a selective breeding programme to fix a qualitative phenotype, he should record the date(s) on which selection occurred, the number of fish that were saved, the number of fish that were culled, and the pond(s) into which the select brood fish were stocked. When the select brood fish spawn, he should record whether they did or did not breed true.

Table: *Example of a data sheet that can be used to record data from a selective breeding programme. In this case, a selective breeding programme that was used to produce a true-breeding population of gold Mozambique tilapia.*

Date: June 1, 1995	
Species: Mozambique tilapia	
Breeding programme: Create a true-breeding gold population	
Phenotype culled	**Number culled**
Black	456
Bronze	935
Phenotype saved	**Number saved**
Gold	204 females and 196 males

Did the population breed true?

Yes, on June 23, 1995, the select brood fish produced only gold offspring

Yes, on July 5, 1995, the select brood fish produced only gold offspring

Yes, on July 17, 1995, the select brood fish produced only gold offspring

If the farmer is going to conduct a progeny test, he should record the date on which each mating was made and the results of each mating. He should record whether he saved or culled a fish and should record where the select brood fish were stocked. When the select brood fish spawn, he should record whether they did or did not breed true. The farmer also must maintain records on where each fish and each family was held until the fish was either saved or culled during the progeny test. Finally, if a farmer is going to produce heterozygotes for grow-out, he should record the ponds which contain each homozygote, the number of fish in each pond, and when they were stocked; the ponds/tanks where the fish are spawned; the number of brood fish that were spawned; and the results of the matings.

Table: *Example of a data sheet that can be used to record data from a progeny test. In this case, a progeny test to cull heterozygous (Rr) red fish and to save homozygous (RR) red fish in order to produce a true-breeding population of red Nile tilapia.*

Dates: May 29–June 15, 1995 Species: Nile tilapia Progeny test: Progeny test red fish to cull heterozygotes and to keep homozygotes. Normally pigmented fish were used as test fish.			
Date	**Red brood fish**	**Progeny ratio (red:normal)**	**Cull/Save**
June 1	1	156:0	Save
June 1	2	73:70	Cull
June 1	3	did not spawn	Cull
June 2	4	75:0	Save
June 2	5	56:61	Cull
Total number of red fish saved: 75 females and 49 males Did the select fish breed true? Yes, on July 25, 1995, the select brood fish produced only red offspring Yes, on July 31, 1995, the select brood fish produced only red offspring			

Table: *Example of a data sheet that can be used to record data about a breeding programme that is used to produce heterozygotes for grow-out. In this case, a breeding programme that mates gold with black fish to produce an all-bronze population of Mozambique tilapia.*

Date: July 25–August 1, 1995		Species: Mozambique tilapia		
Breeding programme: Mating gold × black to produce 100% bronze				
Phenotype	**Date selected**	**Pond No.**	**No. of males**	**No.of females**
Black	June 1, 1995	23	125	230
Gold	June 1, 1995	32	119	198
Production of bronze fish				
Date	**Pond No.**	**Phenotype/sex**	**Phenotype/sex**	**Offspring**
August 1	2	23 gold males	45 black females	8,300 bronze
August 1	3	50 gold females	30 black males	9,500 bronze
August 1	4	45 gold females	32 black males	10,300 bronze

The selective breeding programmes that were not technically difficult and they require minimal record keeping. The breeding programmes needed to produce true-breeding populations are simple and, in most cases, can be conducted in a single breeding season. In some cases, they can be conducted in a single day. Once a true-breeding population is created, selection is finished. Even if a true-breeding population cannot be produced, a simple breeding programme can be designed to produce only the desired phenotype.

The selective breeding programmes that have been used with important aquacultured species to produce populations with only the desired phenotype. For example: The undesired saddleback phenotype (deformed dorsal fin) in the Auburn University strain of blue tilapia was eliminated by a single act of selection, because it was produced by a dominant allele. State fisheries agencies in West Virginia and Pennsylvania (U.S.A.) use both selection and the mating of homozygotes to produce both golden (selection) and palomino (mating of homozygotes) rainbow trout for stocking in public waters. Many common carp farmers in Europe fixed the mirror scale pattern in their populations by culling all other scale phenotypes. Finally, ornamental fish farmers routinely conduct these types of breeding programmes to produce more valuable fish.

Prior to selection, a farmer or farmer cooperative should conduct an economic analysis to determine if a selective breeding programme should be conducted. Two pieces of information are needed: The first is the market values of the phenotypes. The second is the cost of production of the phenotypes, which will be determined by the pleiotropic effects, such as growth rate, survival, etc. If the results indicate that a farmer can increase his profits by fixing a prettier body colour, etc. then he should conduct a selective breeding programme. On the other hand, if the results indicate that increased production costs would exceed the increased market value of the crop, then the farmer should not conduct a selective breeding programme.

A farmer also needs to know when selection is unnecessary because the problem does not need to be corrected. All populations contain deformed individuals. In the wild, fish with deformities are generally eaten by predators, so the percentage of deformed fish that are observed is quite small. Fish farmers exclude predators from their ponds, so fish with deformities are more numerous, even though the incidence is still usually rare. Farmers often think that they have a problem because the odd-looking fish are very noticeable and they remember them. If the frequency of a deformity is less than one in 250 fish, the occurrence is probably not a problem that needs to be solved. On the other hand, if the deformity occurs in one out of every 100–200 fish, the problem should be investigated, and selection may be necessary if the deformity is heritable.

Finally, when a farmer conducts a selective breeding programme, he should save 100–200 select brood fish. The reason for saving this many fish is that a farmer should spawn at least 25 males and 25 females every breeding season. If this is done, he can prevent inbreeding from reaching levels that cause problems.

9

Risks Associated with Transgenic Fish

Food Safety Issues

Potential human consumption of transgenic fish and shellfish raises an assortment of food safety concerns.

- Genes inserted to promote disease resistance may cause transgenic fish to absorb toxic substances (like mercury) at a higher rate and pass these toxic substances on to consumers.
- Roughly 90 percent of food allergies can be attributed to consumption of eggs, fish, shellfish, milk, peanuts, soybeans, tree nuts, and wheat. If proteins used in the production of transgenic species originate from one of these eight sources, there may be potential for allergic reactions among consumers.
- The majority of transgenic fish have been inserted with growth genes. Large doses of growth hormones may pose health risks if consumed in raw and uncooked foods like sushi.

Environmental Concerns

Transgenic fish pose potential threats to natural ecosystems and native species populations that are not fully understood and remain insufficiently studied. However, it is known that:

- Fifty percent of all intentionally introduced fish have had harmful economic or environmental effects;
- Sixty-six percent of all unintentionally introduced fish have had harmful economic or environmental effects;
- Millions of farmed fish escape from open water facilities each year and contaminate native populations; and

- It is inevitable that transgenic fish will escape from aquaculture pens or field trial parameters.

Although these statistics reflect impacts from the releases of non-transgenic fish, the implication is clear – introduction of foreign species to an ecosystem does not occur without negative consequences. The environmental risks of releasing transgenic fish may threaten biodiversity in natural ecosystems and the genetic integrity of those systems.

Genetic alterations in transgenic fish may give them competitive advantages over native species.

- By using growth hormone genes, researchers have been able to increase growth rates 2 to 11 times faster than the normal rate. Faster development leads to earlier sexual maturity and potentially more breeding opportunities than their native counterparts.
- If transgenic fish are genetically enabled to breed earlier and at a faster rate, transgenic genes are more likely to be spread throughout native populations. This would reduce the genetic diversity of the native population.
- Research indicates that transgenic fish may reach maturity faster, but they also die sooner. This "Trojan gene" scenario could have devastating consequences if transgenic fish interbreed with native populations. Additionally, if transgenic genes for rapid growth are correlated with shorter life expectancy, the overall life expectancy for the entire population may be reduced.
- Transgenic fish may have similar effects on natural ecosystems as exotic species. An increased growth rate is often accompanied by a voracious appetite, and transgenic fish may out-compete native species for resources, destroy plants and sensitive habitat, and/or alter the food chain in an ecosystem.

Worldwide Projects

Dozens of genetically engineered fish and shellfish projects are currently underway throughout the world. Genetic engineering is altering the genetics of a variety of species for the purposes of increasing aquaculture production, medical and other research, cleaning up water pollutants, and ornamental reasons. Examples of projects include:

- Inserting genes for increased growth rates and increased resistance to disease in Rainbow trout, tilapia, catfish, grass carp, mollusks, and prawns. These genetic modifications are intended to increase aquaculture productivity for human food consumption.

- Inserting human clotting genes into Tilapia for possible pharmaceutical application.
- Inserting the "anti-freeze" gene into goldfish for research purposes and to add consumer appeal as household pets.
- Algae are being genetically engineered for use as a bioremediation tool. Genetic modification is increasing the algae's ability to absorb higher levels of toxic heavy metals.

For a list of various aquatic transgenic species projects, visit the Pew Initiative on Food and Biotechnology.

States Can Lead the Way

Although there are no federal regulations for transgenic fish, states can protect the health of their citizens and environment. Many states have existing policies in place that could be modified to regulate transgenic fish:

- Fifteen states have adopted regulations concerning uncontained uses of transgenic fish and other transgenic marine organisms.
- California requires permits for genetically engineered species. One year after its enactment, the state had issued fourteen permits, mostly to universities and research institutes.
- Five states have specifically defined genetically engineered organisms as exotic species.
- The Chesapeake Bay Program – a partnership consisting of five states, the District of Columbia, and the Environmental Protection Agency – defines genetically engineered organisms as exotic species.
- The National Institute of Health created research guidelines for genetic engineering. Some states (and some townships within states) adapted their statutes to control genetic research within their borders.
- States could include permit requirements for transgenic fish within existing aquaculture regulations.
- States could include permit requirements for transgenic fish within their Coastal Management Plan.

Genetically Engineered Fish

GM (Genetically Modified) and GMO (Genetically Modified Organisms) have become buzzwords in recent years. Is artificially changing the characteristics of an animal or fish at the genetic level beneficial or an abomination? The debate will rage on for a long time to come. What do you think about genetically engineered fish?.

Selective Breeding

For centuries genetics of animals and fish have been tampered with via selective breeding to enhance desired traits. For nearly two decades the genetic modification of plants, such as soybeans and corn, has resulted in improved crop yields. However, it has only been recently that altering genetic material of fish has taken place

Transgenic Fish

Transgenic fish first hit the ornamental fish market in the form of the GloFish®. The GloFish® is a Zebra Danio that has been given genetic material from jellyfish. The result is a glow in the dark version of one of the most popular aquarium fish. More recently, glow in the dark Convict Cichlids have been created. It's quite likely more will follow.

GE Food

Genetically altered ornamental fish aren't the only ones being created. Transgenic food fish are soon to be on the market. As of late summer 2010, the FDA is poised to approve the first GE (genetically engineered) animal for human consumption. The GE salmon may soon be on our dinner plates.

Safety

In addition to the questions about the safety of eating GE fish, concerns about the environment have been raised. It is quite common for food and ornamental fish to be introduced into local waterways. What effect does an altered fish have on the environment? I don't believe it is likely to be good.

Genetically Modified Foods

A variety of foods can be genetically modified using biotechnology – these are known as GM foods. The genetic material may be altered with methods that do not occur naturally - this is known as 'genetic engineering'. Selected individual genes with specific traits are transferred from one organism to another. Traditional breeding can achieve similar effects, but over a much longer time span. However, traditional breeding cannot achieve the same effects using a transferred gene from a non-related species - this is possible with GM foods.

Genetic Engineering and Plants

Genetic modification of food is not new. For centuries, food crops and animals have been altered through selective breeding. However, while genes can be transferred during selective plant breeding, the scope for exchanging genetic material is much wider using genetic engineering. In theory, genetic engineering allows genetic material to be transferred between any organism, including between plants and animals. For example, the gene from a fish that lives in very cold seas has been inserted into a strawberry, allowing the fruit to be frost-tolerant. This has not as yet been done for currently available commercial food crops. Concerns with climate change may also lead to consideration of GM food crops that are drought tolerant.

Foods that have been Modified

Some foods have been modified to make them resistant to insects and viruses and more able to tolerate herbicides. Crops that have been modified for these purposes in a number of countries, with approval from the relevant authorities, include:

- Maize (corn)
- Wheat
- Rice
- Oilseed rape (canola)
- Chicory
- Squash
- Potato
- Soybean
- Alfalfa
- Cotton.

GM Products in Food

Modified genes may be present in whole foods, such as wheat, soybeans, maize and tomatoes. These GM whole foods are not presently available in Australia. Genetically modified food ingredients **are**, however, present in some Australian foods. For example, soy flour in bread may have come from imported GM soybeans.

Modified genes may have been used in an early stage of the food chain, but may or may not be present in the end product. Gene products - for example, phytochemicals (plant chemicals that contain compounds which may prevent disease) - may, however, remain in the food chain. From a health perspective, this could be an advantage or a disadvantage.

Organic Foods are not Genetically Modified

Foods certified as organic or biodynamic should not contain any GM ingredients, according to industry guidelines.

Nutritional Enhancement

Genetic engineering can also be used to increase the amount of particular nutrients (like vitamins) in food crops. Research into this technique, sometimes called 'nutritional enhancement', is now at an advanced stage. For example, GM golden rice is an example of a white rice crop that has had the vitamin A gene from a daffodil plant inserted. This changes the colour and the vitamin level for countries where vitamin A deficiency is prevalent. Researchers are especially looking at major health problems like iron deficiency. The removal of the proteins that cause allergies from nuts (such as peanuts and Brazil nuts) is also being researched.

Benefits of GM Foods

There is a need to produce inexpensive, safe and nutritious foods to help feed the world's growing population. Genetic modification may provide:

- Sturdy plants able to withstand weather extremes
- Better quality food crops
- Higher nutritional yields in crops
- Inexpensive and nutritious food, like carrots with more antioxidants
- Foods with a greater shelf life, like tomatoes that taste better and last longer
- Food with medicinal (nutraceutical) benefits, such as edible vaccines – for example, bananas with bacterial or rotavirus antigens

- Crops resistant to disease and insects and produce that requires less chemical application, such as pesticide and herbicide resistant plants: for example, GM canola.

GM advocates argue that genetically modified foods are potentially better for the environment. By using genetically engineered crops that are resistant to attack by pests or disease (insect resistant or IR), farmers and primary producers do not have to apply large amounts of pesticides and chemicals to the surrounding environment. Developing crops that are tolerant to particular herbicides (herbicide tolerant or HT) and pesticides may reduce the amount of pesticides used in food production and the residual pesticide levels in the environment.

The Risks of Genetically Modified Crops

Some concerns that have been raised by scientists, community groups and members of the public include:

- New allergens could be inadvertently created - known allergens could be transferred from traditional foods into GM foods. For instance, during laboratory testing, a gene from the Brazil nut was introduced into soybeans. It was found that people with allergies to Brazil nuts could also be allergic to soybeans that had been genetically modified in this way and so the project was ceased. No allergic effects have been found with currently approved GM foods.
- Antibiotic resistance may develop - bioengineers sometimes insert a selectable 'marker' gene to help them identify whether a new gene has been successfully introduced to the host DNA. One such marker gene is for resistance to particular antibiotics. If genes coded for such resistance enter the food chain and are taken up by human gut microflora, the effectiveness of antibiotics could be reduced and human infectious disease risk increased. Research has shown that the risk is very low; however, there is general agreement that use of these markers should be phased out.
- Cross-breeding - other risks include the potential for cross-breeding between GM crops and surrounding vegetation, including weeds. This could result in weeds that are resistant to herbicides and would thus require a greater use of herbicides, which could lead to soil and water contamination. The environmental safety aspects of GM crops vary considerably according to local conditions.

- Herbicide tolerant (HR) crops - the introduction of the glyphosate resistant soybean in 1996 was the start of crops that gave farmers an opportunity to reduce the cost of their herbicide use. However, the increasing acreage of HR crops (such as soybean and canola) has resulted in an increase in the types of weeds that are now glyphosate resistant (GR). These GR weeds may have a major environmental influence on crop production in years to come.
- Pesticide resistant insects - the genetic modification of some crops to permanently produce the natural biopesticide Bacillus thuringiensis (Bt) toxin could encourage the evolution of Bt-resistant insects, rendering the spray ineffective. Wherever pesticides are used, insect resistance can occur and good agricultural practice includes strategies to minimise this.
- Biodiversity - growing GM crops on a large scale may also have implications for biodiversity, the balance of wildlife and the environment. This is why environmental agencies closely monitor their use. Since bees are used to pollinate crops, there is also some suggestion that GM crops may affect organic farming.
- Cross-contamination - plants bioengineered to produce pharmaceuticals (such as medicines) may contaminate food crops. Provisions have been introduced in the USA requiring substantial buffer zones, use of separate equipment and a rule that land used for such crops lie fallow for the next year.
- Pesticide use - the use of pesticide resistant (Bt) crops would suggest a reduction in the application of pesticides; however, recent surveys in the USA suggest that Bt-corn that targets corn borer has not lowered pesticide use, since most pesticides are directed against other corn pests.
- Health effects - minimal research has been conducted into the potential acute or chronic health risks of using GM foods and of their performance in relation to a range of health effects. Research also needs to involve independent (not company-based) assessment of the long-term effects of GM crops in the field and on human health.

Social and Ethical Concerns

Concerns about the social and ethical issues surrounding genetic modification include:

- The possible monopolisation of the world food market by large multinational companies that control the distribution of GM seeds.
- Using genes from animals in plant foods may pose ethical, philosophical or religious problems. For example, eating traces of genetic material from pork could be a problem for certain religious or cultural groups.
- Animal welfare could be adversely affected. For example, cows given more potent GM growth hormones could suffer from health problems related to growth or metabolism.
- New GM organisms could be patented so that 'life' itself could become commercial property through patenting.

Regulation of GM Foods

Current food regulations in Australia state that a GM food will only be approved for sale if it is safe and is as nutritious as its conventional counterparts. Food regulatory authorities require that GM foods receive individual pre-market safety assessments prior to use in foods for human consumption. The principle of 'substantial equivalence' is also used. This means that an existing food is compared with its genetically modified counterpart to find any differences between the existing food and the new product. The assessment investigates:

- Nutritional content
- Toxicity (using similar methods to those used for conventional foods)
- Tendency to provoke any allergic reaction
- Stability of the inserted gene
- Whether there is any nutritional deficit or change in the GM food
- Any other unintended effects of the gene insertion.

The safety of GM foods is still being debated, as it is impossible to predict all of the potential effects on human health and the environment. Some public health experts, however, advocate caution. They believe that we are at the 'scientific starting line' and that we don't know whether GM foods are safe or not.

GM Labelling and the Law

Since December 2002, the law in Australia requires that food labels must show if food has been genetically modified or contains genetically modified ingredients, or whether GM additives or processing aids remain in the final food product. In Australia, GM foods are regulated by the Australia New Zealand Food Standards (FSANZ) Code under Standard 1.5.2 - Food produced using Gene Technology.

Special labels are not required for:

- 'Highly refined' foods where the altered DNA or protein is no longer in the food (for example, oil from modified corn)
- GM food additives or processing aids - unless the new DNA remains in the food to which it is added
- GM flavours where less than 0.1 per cent is present in the food
- Food, food ingredients or processing aids where GM ingredients are 'unintentionally' present in less than 1.0 per cent
- Food that is prepared at the point of sale (takeaway and restaurant food does not have to be labelled).

Labels may be required where:

- Genetic modification has altered the food so that its composition or nutritional value is 'outside the normal range' of similar non-GM goods; for example, if GM technology is used to add vitamins or omega-3 fatty acids
- Naturally occurring toxins are 'significantly different' to similar non-GM foods
- The food produced using GM technology contains a 'new factor', which can cause allergic reactions in some people
- Genetic modification raises 'significant ethical, cultural and religious concerns' regarding the origin of the genetic material used.

GM Food on the Shelves

Many foods on supermarket shelves contain imported GM ingredients. A variety of GM foods have also been approved for production in Australia. These foods include corn, soybeans, potatoes and canola. Others are still undergoing field trials approved by the Office of the Gene Technology Regulator (OGTR), although the moratorium by State Governments (lifted in Victoria and NSW in early 2008) stopped some GM field trials. Imported food products are subject to the same regulations as domestically manufactured foods.

There are around 20 GM foods, additives, flavourings, growth hormone (bovine somatotropin) and enzymes (like rennet, used to make cheese) currently approved in Europe. In the USA, there are more than 40 approved GM foods. The main sources of GM foods in Australia include:

- Imported soya from the United States - this is one of the main sources of GM ingredients in food sold in Australia since 1996. The soya has been genetically modified to be resistant to a herbicide. It can be found in a wide range of foods, such as chocolates, potato chips, margarine, mayonnaise, biscuits and bread.
- Cottonseed oil made from GM cotton - this oil, made from cotton that is resistant to a pesticide, is used in Australia for frying (by the food industry) and in mayonnaise and salad dressings.
- Imported GM corn - this is mainly used as cattle feed at present and has not been approved for farming in Australia. However, GM corn may have entered the Australian market through imported foods like breakfast cereal, bread, corn chips and gravy mixes. If so, it is now required to be labelled.
- Other GM foods available overseas - these may be ingredients in foods imported to Australia including potatoes, canola oil, sugar beet, yeast, cauliflower and coffee.

If you Want GM-Free Food

Due to consumer demand, some food manufacturers in Australia have taken steps to provide GM-free food. These products may be labelled accordingly; for example, 'contains no genetically modified ingredients'. Although Food Standards Australia New Zealand (FSANZ) does not provide a consumer hotline on GM matters, people can make enquiries to the Office of the Gene Technology Regulator.

Problems Associated with Salmon Farming

Environmental Media Services raises its concerns and opinions on the impact of Salmon Farming in the Environment.

There are a number of environmental issues associated with salmon aquaculture:

Interactions with Wild Fish: When fish escape from farms and survive in large numbers or establish their own breeding populations, they will compete with wild salmon. If they are the same species as the wild salmon (i.e. Atlantic salmon grown in the Atlantic), there is the possibility of interbreeding between farmed and wild fish. When such interbreeding occurs, there is a significant change in the genetics of the salmon population. More.

Genetically Engineered Salmon: The introduction of genetically altered salmon in ocean pen farms — a distinct possibility in the near future — adds another layer of concern with respect to interactions with wild populations. If interbreeding were to occur as a result of escapes, such genes could be incorporated into the wild gene pool and possibly diminish the vigour of the wild population. Hatcheries are problematic because the selective survival of large numbers of young from the small numbers of adults that donate eggs and sperm change the genetic pool of the wild population. More.

Wild Fish for Feed: Farmed salmon are fed meal and oils from wild-caught fish. Each pound of salmon produced requires at least 3 pounds of wild-caught fish, challenging the presumption that fish farming necessarily reduces commercial fishing pressure. In fact, there is a net loss of protein in the marine ecosystem as a whole when wild catch is converted into meal for aquaculture consumption.

Pollution: Pens full of salmon produce large amounts of waste — both excrement and unconsumed feed. This may result in water quality conditions (such as high nutrients and low oxygen) that are unfavourable for both the farmed fish and the natural ecosystem. It is suspected that nutrients released from salmon farms stimulate micro- algal blooms, but proof of this is lacking because little research has been done.

Disease: The densely packed condition in pens promotes disease, a common problem in most salmon farms. Furthermore, there have been documented disease transfers from farmed salmon to wild populations, and the potential effects are serious. While antibiotics are used to treat some diseases, there are concerns about the effects of antibiotic-resistant bacteria on human health. There has been an emphasis on developing vaccines to prevent specific diseases, which reduces the need for antibiotics. More.

Aesthetics: In some areas, landowners have opposed salmon aquaculture facilities sited near residential shorelines because they are unsightly, odoriferous and/or interfere with the natural setting of the seascape. In fact, in the U.S., aquaculture development has been significantly slowed down by these concerns.

The Causes

Escapes are inevitable in open-water pens, especially during storm conditions. Some incidents have resulted in the escape of tens of thousands of farmed fish into the environment. Efforts to secure facilities against these accidents may reduce the size and number of releases, but is unlikely to stop them altogether. The density of salmon in farms is variable, but the farmer is motivated to pack them at high densities to increase profits. This exacerbates the problems of pollution and disease and causes stress on the fish that leads to inferior product quality.

The siting of salmon farms is often problematic, particularly if it does not adequately take into account the proximity to wild salmon migration routes, water flow and circulation patterns, the fate of waste materials, the number of facilities already in an area and aesthetic concerns. According to some, the development of genetically engineered salmon for farming is proceeding without concern for the consequences of inevitable escapes into coastal waters inhabited by wild salmon.

The Context

Populations of salmon in the western North Atlantic have recently plummeted for unknown reasons. Meanwhile, in the eastern Pacific, more than 100 populations have disappeared and salmon are extinct in 40 percent of the rivers where they once spawned along the North American Pacific coast. The potential for interactions with farmed fish and transmission of disease from farmed to wild salmon is especially threatening in the context of these declines.

Governments typically encourage aquaculture because it is viewed as economic development. This often leads to the intensive, large-scale farming methods most often associated with environmental damage.

The market price of farmed salmon is artificially low. Because the costs of environmental damage are not borne by the industry, nor are the value of ecosystem services factored into the cost of production, there is no pressure on the industry to operate in environmentally sound ways.

Genetically Modified Foods: Harmful or Helpful

Genetically-modified foods (GM foods) have made a big splash in the news lately. European environmental organisations and public interest groups have been actively protesting against GM foods for months, and recent controversial studies about the effects of genetically-modified corn pollen on monarch butterfly caterpillars have brought the issue of genetic engineering to the forefront of the public consciousness in the U.S. In response to the upswelling of public concern, the U.S. Food and Drug Administration (FDA) held three open meetings in Chicago, Washington, D.C., and Oakland, California to solicit public opinions and begin the process of establishing a new regulatory procedure for government approval of GM foods. I attended the FDA meeting held in November 1999 in Washington, D.C., and here I will attempt to summarise the issues involved and explain the U.S. government's present role in regulating GM food.

What are Genetically-Modified Foods?

The term GM foods or GMOs (genetically-modified organisms) is most commonly used to refer to crop plants created for human or animal consumption using the latest molecular biology techniques. These plants have been modified in the laboratory to enhance desired traits such as increased resistance to herbicides or improved nutritional content. The enhancement of desired traits has traditionally been undertaken through breeding, but conventional plant breeding methods can be very time consuming and are often not very accurate.

Genetic engineering, on the other hand, can create plants with the exact desired trait very rapidly and with great accuracy. For example, plant geneticists can isolate a gene responsible for drought tolerance and insert that gene into a different plant. The new genetically-modified plant will gain drought tolerance as well. Not only can genes be transferred from one plant to another, but genes from non-plant organisms also can be used. The best known example of this is the use of B.t. genes in corn and other crops. B.t., or *Bacillus thuringiensis*, is a naturally occurring bacterium that produces crystal proteins that are lethal to insect larvae. B.t. crystal protein genes have been transferred into corn, enabling the corn to produce its own pesticides against insects such as the European corn borer. For two informative overviews of some of the techniques involved in creating GM foods, visit Biotech Basics (sponsored by

Monsanto) or Techniques of Plant Biotechnology from the National Centre for Biotechnology Education.

What are Some of the Advantages of GM Foods?

The world population has topped 6 billion people and is predicted to double in the next 50 years. Ensuring an adequate food supply for this booming population is going to be a major challenge in the years to come. GM foods promise to meet this need in a number of ways:

- *Pest Resistance:* Crop losses from insect pests can be staggering, resulting in devastating financial loss for farmers and starvation in developing countries. Farmers typically use many tons of chemical pesticides annually. Consumers do not wish to eat food that has been treated with pesticides because of potential health hazards, and run-off of agricultural wastes from excessive use of pesticides and fertilizers can poison the water supply and cause harm to the environment. Growing GM foods such as B.t. corn can help eliminate the application of chemical pesticides and reduce the cost of bringing a crop to market.
- *Herbicide Tolerance:* For some crops, it is not cost-effective to remove weeds by physical means such as tilling, so farmers will often spray large quantities of different herbicides (weed-killer) to destroy weeds, a time-consuming and expensive process, that requires care so that the herbicide doesn't harm the crop plant or the environment. Crop plants genetically-engineered to be resistant to one very powerful herbicide could help prevent environmental damage by reducing the amount of herbicides needed. For example, Monsanto has created a strain of soybeans genetically modified to be not affected by their herbicide product Roundup ®. A farmer grows these soybeans which then only require one application of weed-killer instead of multiple applications, reducing production cost and limiting the dangers of agricultural waste run-off.
- *Disease Resistance:* There are many viruses, fungi and bacteria that cause plant diseases. Plant biologists are working to create plants with genetically-engineered resistance to these diseases.
- *Cold Tolerance:* Unexpected frost can destroy sensitive seedlings. An antifreeze gene from cold water fish has been introduced into plants such as tobacco and potato. With this antifreeze gene, these plants are able to tolerate cold

temperatures that normally would kill unmodified seedlings. (Note: I have not been able to find any journal articles or patents that involve fish antifreeze proteins in strawberries, although I have seen such reports in newspapers. I can only conclude that nothing on this application has yet been published or patented.)

- *Drought tolerance/salinity Tolerance:* As the world population grows and more land is utilised for housing instead of food production, farmers will need to grow crops in locations previously unsuited for plant cultivation. Creating plants that can withstand long periods of drought or high salt content in soil and groundwater will help people to grow crops in formerly inhospitable places.
- *Nutrition:* Malnutrition is common in third world countries where impoverished peoples rely on a single crop such as rice for the main staple of their diet. However, rice does not contain adequate amounts of all necessary nutrients to prevent malnutrition. If rice could be genetically engineered to contain additional vitamins and minerals, nutrient deficiencies could be alleviated. For example, blindness due to vitamin A deficiency is a common problem in third world countries. Researchers at the Swiss Federal Institute of Technology Institute for Plant Sciences have created a strain of "golden" rice containing an unusually high content of beta-carotene (vitamin A). Since this rice was funded by the Rockefeller Foundation, a non-profit organisation, the Institute hopes to offer the golden rice seed free to any third world country that requests it. Plans were underway to develop a golden rice that also has increased iron content. However, the grant that funded the creation of these two rice strains was not renewed, perhaps because of the vigorous anti-GM food protesting in Europe, and so this nutritionally-enhanced rice may not come to market at all.
- *Pharmaceuticals:* Medicines and vaccines often are costly to produce and sometimes require special storage conditions not readily available in third world countries. Researchers are working to develop edible vaccines in tomatoes and potatoes. These vaccines will be much easier to ship, store and administer than traditional injectable vaccines.

- *Phytoremediation:* Not all GM plants are grown as crops. Soil and groundwater pollution continues to be a problem in all parts of the world. Plants such as poplar trees have been genetically engineered to clean up heavy metal pollution from contaminated soil.

How Prevalent are GM Crops?

What Plants are Involved?

According to the FDA and the United States Department of Agriculture (USDA), there are over 40 plant varieties that have completed all of the federal requirements for commercialisation. Some examples of these plants include tomatoes and cantalopes that have modified ripening characteristics, soybeans and sugarbeets that are resistant to herbicides, and corn and cotton plants with increased resistance to insect pests. Not all these products are available in supermarkets yet; however, the prevalence of GM foods in U.S. grocery stores is more widespread than is commonly thought. While there are very, very few genetically-modified whole fruits and vegetables available on produce stands, highly processed foods, such as vegetable oils or breakfast cereals, most likely contain some tiny percentage of genetically-modified ingredients because the raw ingredients have been pooled into one processing stream from many different sources. Also, the ubiquity of soybean derivatives as food additives in the modern American diet virtually ensures that all U.S. consumers have been exposed to GM food products.

The U.S. statistics that follow are derived from data presented on the USDA. The global statistics are derived from a brief published by the International Service for the Acquisition of Agri-biotech Applications (ISAAA), and from the Biotechnology Industry Organisation.

Thirteen countries grew genetically-engineered crops commercially in 2000, and of these, the U.S. produced the majority. In 2000, 68% of all GM crops were grown by U.S. farmers. In comparison, Argentina, Canada and China produced only 23%, 7% and 1%, respectively. Other countries that grew commercial GM crops in 2000 are Australia, Bulgaria, France, Germany, Mexico, Romania, South Africa, Spain, and Uruguay.

Soybeans and corn are the top two most widely grown crops (82% of all GM crops harvested in 2000), with cotton, rapeseed (or canola) and potatoes trailing behind. 74% of these GM crops were modified for herbicide tolerance, 19% were modified for insect pest resistance, and 7% were modified for both herbicide tolerance and pest tolerance. Globally, acreage of GM crops has increased 25-fold in just 5 years, from approximately 4.3 million acres in 1996 to 109 million acres in 2000 - almost twice the area of the United Kingdom. Approximately 99 million acres were devoted to GM crops in the U.S. and Argentina alone.

In the U.S., approximately 54% of all soybeans cultivated in 2000 were genetically-modified, up from 42% in 1998 and only 7% in 1996. In 2000, genetically-modified cotton varieties accounted for 61% of the total cotton crop, up from 42% in 1998, and 15% in 1996. GM corn and also experienced a similar but less dramatic increase. Corn production increased to 25% of all corn grown in 2000, about the same as 1998 (26%), but up from 1.5% in 1996. As anticipated, pesticide and herbicide use on these GM varieties was slashed and, for the most part, yields were increased.

What are Some of the Criticisms against GM Foods?

Environmental activists, religious organisations, public interest groups, professional associations and other scientists and government officials have all raised concerns about GM foods, and criticized agribusiness for pursuing profit without concern for potential hazards, and the government for failing to exercise adequate regulatory oversight. It seems that everyone has a strong opinion about GM foods. Even the Vatican and the Prince of Wales have expressed their opinions. Most concerns about GM foods fall into three categories: environmental hazards, human health risks, and economic concerns.

Environmental Hazards

- Unintended harm to other organisms Last year a laboratory study was published in Nature showing that pollen from B.t. corn caused high mortality rates in monarch butterfly caterpillars. Monarch caterpillars consume milkweed plants, not corn, but the fear is that if pollen from B.t. corn is blown by the wind onto milkweed plants in neighbouring fields, the caterpillars could eat the pollen and perish. Although the Nature study was not conducted under natural field conditions, the results seemed to support this viewpoint. Unfortunately, B.t. toxins kill many species of insect larvae indiscriminately;

it is not possible to design a B.t. toxin that would only kill crop-damaging pests and remain harmless to all other insects. This study is being reexamined by the USDA, the U.S. Environmental Protection Agency (EPA) and other non-government research groups, and preliminary data from new studies suggests that the original study may have been flawed. This topic is the subject of acrimonious debate, and both sides of the argument are defending their data vigorously. Currently, there is no agreement about the results of these studies, and the potential risk of harm to non-target organisms will need to be evaluated further.

- Reduced effectiveness of pesticides Just as some populations of mosquitoes developed resistance to the now-banned pesticide DDT, many people are concerned that insects will become resistant to B.t. or other crops that have been genetically-modified to produce their own pesticides.
- Gene transfer to non-target species Another concern is that crop plants engineered for herbicide tolerance and weeds will cross-breed, resulting in the transfer of the herbicide resistance genes from the crops into the weeds. These "superweeds" would then be herbicide tolerant as well. Other introduced genes may cross over into non-modified crops planted next to GM crops. The possibility of interbreeding is shown by the defence of farmers against lawsuits filed by Monsanto. The company has filed patent infringement lawsuits against farmers who may have harvested GM crops. Monsanto claims that the farmers obtained Monsanto-licensed GM seeds from an unknown source and did not pay royalties to Monsanto. The farmers claim that their unmodified crops were cross-pollinated from someone else's GM crops planted a field or two away. More investigation is needed to resolve this issue.

There are several possible solutions to the three problems mentioned above. Genes are exchanged between plants via pollen. Two ways to ensure that non-target species will not receive introduced genes from GM plants are to create GM plants that are male sterile (do not produce pollen) or to modify the GM plant so that the pollen does not contain the introduced gene. Cross-pollination would not occur, and if harmless insects such as monarch caterpillars were to eat pollen from GM plants, the caterpillars would survive.

Another possible solution is to create buffer zones around fields of GM crops. For example, non-GM corn would be planted to surround a field of B.t. GM corn, and the non-GM corn would not be harvested. Beneficial or harmless insects would have a refuge in the non-GM corn, and insect pests could be allowed to destroy the non-GM corn and would not develop resistance to B.t. pesticides. Gene transfer to weeds and other crops would not occur because the wind-blown pollen would not travel beyond the buffer zone. Estimates of the necessary width of buffer zones range from 6 metres to 30 metres or more. This planting method may not be feasible if too much acreage is required for the buffer zones.

Human Health Risks

- Allergenicity Many children in the US and Europe have developed life-threatening allergies to peanuts and other foods. There is a possibility that introducing a gene into a plant may create a new allergen or cause an allergic reaction in susceptible individuals. A proposal to incorporate a gene from Brazil nuts into soybeans was abandoned because of the fear of causing unexpected allergic reactions. Extensive testing of GM foods may be required to avoid the possibility of harm to consumers with food allergies. Labelling of GM foods and food products will acquire new importance, which I shall discuss later.
- Unknown effects on human health There is a growing concern that introducing foreign genes into food plants may have an unexpected and negative impact on human health. A recent article published in Lancet examined the effects of GM potatoes on the digestive tract in rats. This study claimed that there were appreciable differences in the intestines of rats fed GM potatoes and rats fed unmodified potatoes. Yet critics say that this paper, like the monarch butterfly data, is flawed and does not hold up to scientific scrutiny. Moreover, the gene introduced into the potatoes was a snowdrop flower lectin, a substance known to be toxic to mammals. The scientists who created this variety of potato chose to use the lectin gene simply to test the methodology, and these potatoes were never intended for human or animal consumption.

On the whole, with the exception of possible allergenicity, scientists believe that GM foods do not present a risk to human health.

Economic Concerns

Bringing a GM food to market is a lengthy and costly process, and of course agri-biotech companies wish to ensure a profitable return on their investment. Many new plant genetic engineering technologies and GM plants have been patented, and patent infringement is a big concern of agribusiness. Yet consumer advocates are worried that patenting these new plant varieties will raise the price of seeds so high that small farmers and third world countries will not be able to afford seeds for GM crops, thus widening the gap between the wealthy and the poor. It is hoped that in a humanitarian gesture, more companies and non-profits will follow the lead of the Rockefeller Foundation and offer their products at reduced cost to impoverished nations.

Patent enforcement may also be difficult, as the contention of the farmers that they involuntarily grew Monsanto-engineered strains when their crops were cross-pollinated shows. One way to combat possible patent infringement is to introduce a "suicide gene" into GM plants. These plants would be viable for only one growing season and would produce sterile seeds that do not germinate. Farmers would need to buy a fresh supply of seeds each year. However, this would be financially disastrous for farmers in third world countries who cannot afford to buy seed each year and traditionally set aside a portion of their harvest to plant in the next growing season. In an open letter to the public, Monsanto has pledged to abandon all research using this suicide gene technology.

How are GM Foods Regulated and what is the Government's Role in this Process?

Governments around the world are hard at work to establish a regulatory process to monitor the effects of and approve new varieties of GM plants. Yet depending on the political, social and economic climate within a region or country, different governments are responding in different ways.

In Japan, the Ministry of Health and Welfare has announced that health testing of GM foods will be mandatory as of April 2001. Currently, testing of GM foods is voluntary. Japanese supermarkets are offering both GM foods and unmodified foods, and customers are beginning to show a strong preference for unmodified fruits and vegetables.

India's government has not yet announced a policy on GM foods because no GM crops are grown in India and no products are commercially available in supermarkets yet. India is, however, very supportive of transgenic plant research. It is highly likely that India will decide that the benefits of GM foods outweigh the risks because Indian agriculture will need to adopt drastic new measures to counteract the country's endemic poverty and feed its exploding population.

Some states in Brazil have banned GM crops entirely, and the Brazilian Institute for the Defence of Consumers, in collaboration with Greenpeace, has filed suit to prevent the importation of GM crops. Brazilian farmers, however, have resorted to smuggling GM soybean seeds into the country because they fear economic harm if they are unable to compete in the global marketplace with other grain-exporting countries.

In Europe, anti-GM food protestors have been especially active. In the last few years Europe has experienced two major foods scares: bovine spongiform encephalopathy (mad cow disease) in Great Britain and dioxin-tainted foods originating from Belgium. These food scares have undermined consumer confidence about the European food supply, and citizens are disinclined to trust government information about GM foods. In response to the public outcry, Europe now requires mandatory food labelling of GM foods in stores, and the European Commission (EC) has established a 1% threshold for contamination of unmodified foods with GM food products.

In the United States, the regulatory process is confused because there are three different government agencies that have jurisdiction over GM foods. To put it very simply, the EPA evaluates GM plants for environmental safety, the USDA evaluates whether the plant is safe to grow, and the FDA evaluates whether the plant is safe to eat. The EPA is responsible for regulating substances such as pesticides or toxins that may cause harm to the environment. GM crops such as B.t. pesticide-laced corn or herbicide-tolerant crops but not foods modified for their nutritional value fall under the purview of the EPA. The USDA is responsible for GM crops that do not fall under the umbrella of the EPA such as drought-tolerant or disease-tolerant crops, crops grown for animal feeds, or whole fruits, vegetables and grains for human consumption. The FDA historically has been concerned with pharmaceuticals, cosmetics and food products and additives, not whole foods. Under current guidelines, a genetically-modified ear of corn sold at a produce stand is not regulated by the FDA because it is a whole food, but a box of cornflakes is regulated because it is a food product. The FDA's stance is that GM foods are substantially equivalent to unmodified, "natural" foods, and therefore not subject to FDA regulation.

The EPA conducts risk assessment studies on pesticides that could potentially cause harm to human health and the environment, and establishes tolerance and residue levels for pesticides. There are strict limits on the amount of pesticides that may be applied to crops during growth and production, as well as the amount that remains in the food after processing. Growers using pesticides must have a license for each pesticide and must follow the directions on the label to accord with the EPA's safety standards. Government inspectors may periodically visit farms and conduct investigations to ensure compliance. Violation of government regulations may result in steep fines, loss of license and even jail sentences.

As an example the EPA regulatory approach, consider B.t. corn. The EPA has not established limits on residue levels in B.t corn because the B.t. in the corn is not sprayed as a chemical pesticide but is a gene that is integrated into the genetic material of the corn itself. Growers must have a license from the EPA for B.t corn, and the EPA has issued a letter for the 2000 growing season requiring farmers to plant 20% unmodified corn, and up to 50% unmodified corn in regions where cotton is also cultivated. This planting strategy may help prevent insects from developing resistance to the B.t. pesticides as well as provide a refuge for non-target insects such as Monarch butterflies.

The USDA has many internal divisions that share responsibility for assessing GM foods. Among these divisions are APHIS, the Animal Health and Plant Inspection Service, which conducts field tests and issues permits to grow GM crops, the Agricultural Research Service which performs in-house GM food research, and the Cooperative State Research, Education and Extension Service which oversees the USDA risk assessment program. The USDA is concerned with potential hazards of the plant itself. Does it harbor insect pests? Is it a noxious weed? Will it cause harm to indigenous species if it escapes from farmer's fields? The USDA has the power to impose quarantines on problem regions to prevent movement of suspected plants, restrict import or export of suspected plants, and can even destroy plants cultivated in violation of USDA regulations. Many GM plants do not require USDA permits from APHIS. A GM plant does not require a permit if it meets these 6 criteria: 1) the plant is not a noxious weed; 2) the genetic material introduced into the GM plant is stably integrated into the plant's own genome; 3) the function of the introduced gene is known and does not cause plant disease; 4) the GM plant is not toxic to non-target organisms; 5) the introduced gene will not cause the creation of new plant viruses; and 6) the GM plant cannot contain genetic material from animal or human pathogens.

The current FDA policy was developed in 1992 (Federal Register Docket No. 92N-0139) and states that agri-biotech companies may voluntarily ask the FDA for a consultation. Companies working to create new GM foods are not required to consult the FDA, nor are they required to follow the FDA's recommendations after the consultation. Consumer interest groups wish this process to be mandatory, so that all GM food products, whole foods or otherwise, must be approved by the FDA before being released for commercialisation. The FDA counters that the agency currently does not have the time, money, or resources to carry out exhaustive health and safety studies of every proposed GM food product. Moreover, the FDA policy as it exists today does not allow for this type of intervention.

How are GM Foods Labelled?

Labelling of GM foods and food products is also a contentious issue. On the whole, agribusiness industries believe that labelling should be voluntary and influenced by the demands of the free market. If consumers show preference for labelled foods over non-labelled foods, then industry will have the incentive to regulate itself or risk alienating the customer. Consumer interest groups, on the other hand, are demanding mandatory labelling. People have the right to know what they are eating, argue the interest groups, and historically industry has proven itself to be unreliable at self-compliance with existing safety regulations. The FDA's current position on food labelling is governed by the Food, Drug and Cosmetic Act which is only concerned with food additives, not whole foods or food products that are considered "GRAS" - generally recognised as safe. The FDA contends that GM foods are substantially equivalent to non-GM foods, and therefore not subject to more stringent labelling. If all GM foods and food products are to be labelled, Congress must enact sweeping changes in the existing food labelling policy.

There are many questions that must be answered if labelling of GM foods becomes mandatory. First, are consumers willing to absorb the cost of such an initiative? If the food production industry is required to label GM foods, factories will need to construct two separate processing streams and monitor the production lines accordingly. Farmers must be able to keep GM crops and non-GM crops from mixing during planting, harvesting and shipping. It is almost assured that industry will pass along these additional costs to consumers in the form of higher prices.

Secondly, what are the acceptable limits of GM contamination in non-GM products? The EC has determined that 1% is an acceptable limit of cross-contamination, yet many consumer interest groups argue that only 0% is acceptable. Some companies such as Gerber baby foods and Frito-Lay have pledged to avoid use of GM foods in any of their products. But who is going to monitor these companies for compliance and what is the penalty if they fail? Once again, the FDA does not have the resources to carry out testing to ensure compliance.

What is the level of detectability of GM food cross-contamination? Scientists agree that current technology is unable to detect minute quantities of contamination, so ensuring 0% contamination using existing methodologies is not guaranteed. Yet researchers disagree on what level of contamination really is detectable, especially in highly processed food products such as vegetable oils or breakfast cereals where the vegetables used to make these products have been pooled from many different sources. A 1% threshold may already be below current levels of detectability.

Finally, who is to be responsible for educating the public about GM food labels and how costly will that education be? Food labels must be designed to clearly convey accurate information about the product in simple language that everyone can understand. This may be the greatest challenge faced be a new food labelling policy: how to educate and inform the public without damaging the public trust and causing alarm or fear of GM food products.

In January 2000, an international trade agreement for labelling GM foods was established. More than 130 countries, including the US, the world's largest producer of GM foods, signed the agreement. The policy states that exporters must be required to label all GM foods and that importing countries have the right to judge for themselves the potential risks and reject GM foods, if they so choose. This new agreement may spur the U.S. government to resolve the domestic food labelling dilemma more rapidly.

Genetically-modified foods have the potential to solve many of the world's hunger and malnutrition problems, and to help protect and preserve the environment by increasing yield and reducing reliance upon chemical pesticides and herbicides. Yet there are many challenges ahead for governments, especially in the areas of safety testing, regulation, international policy and food labelling. Many people feel that genetic engineering is the inevitable wave of the future and that we cannot afford to ignore a technology that has such enormous potential benefits. However, we must proceed with caution to avoid causing unintended harm to human health and the environment as a result of our enthusiasm for this powerful technology.

Genetically Modified Salmon: Changing The Future

Twenty years of research has meant that consumers are closer than ever to seeing genetically modified (GM) salmon on their plates. Charlotte Johnston, The FishSite junior editor speaks with Dr Ronald Stotish, CEO and President of AquaBounty Technologies, the small company that wants to make a difference.

Producing market weight salmon in half the time of conventional salmon, is what AquaBounty are claiming to be able to do.

The AquAdvantage salmon is transgenic, or genetically modified. Whilst there are plenty of GM plants, which are sold on the US and global markets, there are no GM animal food products available.

Approval must be sought from the US Food and Drug Administration (FDA) before the product is sold. AquaBounty sent in an original application in 1995, however, it was not until 2009 that the company had completed all outstanding FDA submissions and requests for information.

Now they are sitting, waiting for approval.

How Does it Work?

Naturally salmon are slow growers, in the first year of life they may achieve a weight of between 20-30 grams. Transgenic salmon could change that.

Dr Stotish explained that taking the growth hormone from Chinook salmon, and putting it into fertilised Atlantic salmon eggs, allows the Atlantic salmon to grow more like a trout, reaching desired market weight in a much faster time.

The Chinook salmon's growth promoting gene allows it to grow in less hospitable environments. By transferring this gene, which uses the same growth promoting hormone in both species, AquAdvantage salmon has the ability to grow in warmer waters.

The final product is AquAdvantage salmon, all of which are sterile females.

Figure: *AquAdvantage salmon and Atlantic salmon at the same age*

The Benefits

With AquAdvantage salmon, market weight of around 2-3kgs can be achieved in 18 - 24 months, as opposed to three years.

However faster growth is not the only benefit, says Dr Stotish.

AquAdvantage salmon can be grown domestically, in containment close to land and populations. This provides a number of benefits.

The first would be that producing the fish in captivity, in closed systems, would dramatically reduce transport costs and improve the whole supply chain. Dr Stotish explained that this would have a huge environmental impact, reducing the industry's carbon footprint.

Secondly, producing more salmon in containment will reduce pressure on wild fish stocks. Improved fishing methods and overfishing has resulted in a major decline in many fish stocks. There are huge concerns for wild stocks of Atlantic salmon which can reach closer to 5-6kgs in weight.

Thirdly, allowing production of salmon close to populations will mean that a fresh supply of fish is available at all times. On top of this, Dr Stotish says that with the Food and Agriculture Organisation (FAO) predicting that world food production must increase by 70 per cent between now and 2050, aquaculture could play a major role.

FDA Approval

Saying how the first application to the FDA was in 1995, Dr Stotish says that it wasn't until 2009, that AquaBounty had submitted all the information that the FDA required.

He says that it was also last year that the FDA published a guidance document to regulate transgenic animals, he says it is similar to that which regulates veterinary drugs.

Every case is reviewed completely separately, on risk based status. Food safety, the environment and a number of other factors are considered.

AquaBounty has a vast amount of information on Atlantic salmon, having studied them for the last 20 years.

With all the materials finally submitted to the FDA, he hopes a decision will be reached by the end of the year.

Concerns

A recent article, written by Wenonah Hauter, Chairman of Food and Water Watch, highlighted a number of concerns surrounding the approval of GM salmon.

She raised valid points, questioning environmental issues as well as consumer health.

Ms Hauter was concerned that the AquAdvantage salmon may escape from captivity. This would present two problems: the first that they would breed with wild salmon, and the second that they would compete and breed with wild salmon, and that the GM salmon, if they did escape, would outcompete wild salmon for natural resources.

Dr Stotish said that as all AquAdvantage salmon are sterile females, there is no chance of them breeding with wild salmon.

He also said that AquaBounty has provided AquAdvantage salmon for a three-year study carried out by the University of Newfoundland.

The study, which is coming to an end, looked at the species ability to compete in the wild. Although the conclusions are only preliminary, researcher Dr Eric Hallerman says that he does not believe the AquAdvantage salmon to be able to compete in the wild, as they are domesticated. Preliminary results also say that AquAdvantage salmon struggle when feed is limited, making them less competitive than their close relations, the Atlantic salmon.

Dr Stotish say that a number of other independent studies have confirmed that AquAdvantage salmon are no threat to the environment.

Dr Stotish guaranteed that the end product is 100 per cent safe for the consumer to eat. However, if approved, it will be two years until GM salmon is for sale.

Production

These salmon would be produced in captivity. Dr Stotish says that the breed is slightly more feed efficient, he reckons by about 10 per cent. "With more studying and experience, this could be improved."

Animal health and safety studies have confirmed that are no welfare issues which effect the fish produced from the eggs.

Marketing

The cost, at which the eggs will be marketed, is still unknown. However, Dr Stotish says that due to the faster growth, and therefore improved economics, they will undoubtedly be higher than their Atlantic salmon counterparts.

The other benefits, such as being able to supply fresh seafood, cut transport costs etc will all be factored into the final price.

Without giving too much away, Dr Stotish said that they did have a number potential clients who were interested in purchasing the salmon eggs.

Asked how he thinks the availability of transgenic salmon will affect market prices, Dr Stotish says that this is unknown.

Eventually, Dr Stotish is aware that other companies will start selling similar products. However, for now, he believes that the regulatory hurdle is a barrier to competition.

Has AquaBounty considered what the final consumer wants? Whilst the company has carried out no official research, the International Food Information Council has carried out a number of surveys asking consumers whether they would eat FDA approved GM food.

The most recent of these surveys suggest that the public are becoming more favourable towards the benefits of biotechnology. 41 per cent of respondents were 'very favourable' or 'somewhat favourable' towards genetic modification of animals.

Lack of information appears to be the primary reason why consumers are not favourable towards animal biotechnology.

"There are no differences in flavour, nutritional values or flesh colour. The two fish are undistinguishable," says Dr Stotish, talking about AquAdvantage salmon and Atlantic salmon.

In fact, he pointed out that in one tasting session, AquAdvantage salmon had been referred to as the tastier fish. AquAdvantage salmon is an environmentally friendly product, he said.

Dr Stotish also pointed out that if AquAdvantage salmon was cheaper, it may be more appealing to consumers.

What's Next?

Transgenic salmon is just one of many things that AquaBounty are involved in. They have transferred the technology to other species, looking at faster growing rainbow trout and tilapia - although neither of these are close to becoming approved by the FDA.

The possibilities are vast, says Dr Stotish, explaining how the company is currently looking at developing a technology that treats virus infections in shrimp.

The technology can be used to evolve disease resistance traits, or enhance taste, he continues.

Dr Stotish says he wants to make a sustainable food supply, but at what cost?

If and when finally approved, Aqua Bounty will have paved the way for genetically modified food from animals. Its scope is endless.

The question is, which is more important to the consumer - the benefits of AquAdvantage salmon, or the ideas and phobia that exist surrounding GM food?

Careful Risk Assessment Needed to Evaluate Transgenic Fish

Abstract

The reproductive biology of fish makes them particularly amenable to genetic manipulation. A genetically engineered or "transgenic"

Atlantic salmon is currently undergoing federal regulatory review, and international research is being conducted on many other species. The innate ability of fish to escape confinement and potentially invade native ecosystems elevates the ecological concerns associated with their genetic modification. Escaped transgenic fish will not invariably result in deleterious effects on native populations, and careful risk assessment is required to determine the ecological risks unique to each transgene, species and receiving ecosystem combination. In response to public concerns about transgenic fish, California has developed stringent regulations for the importation, possession and raising of transgenic fish, and a California law prohibits their presence in waters of the Pacific Ocean regulated by the state.

Full Text

The rationale for developing transgenic or genetically engineered animals for agricultural applications is essentially to increase their productivity and yield, improve their resistance to diseases and parasites, and enhance the nutritional and processing qualities of foods derived from these transgenic animals. Compared with mammals, fish offer important advantages for the production of transgenics because of the large number of eggs laid per female, the fact that fertilization and embryonic development takes place outside the mother (in most species), the lower probability of carrying human pathogens, and the fact that aquaculture is a rapidly expanding market. The first transgenic fish were produced in 1984, and since that time more than 30 species have been genetically engineered worldwide. The number of transgenic species is higher for fish than for all other vertebrate species combined. Transgenic fish have been developed for applications such as the production of human therapeutics, experimental models for biological research, environmental monitoring, ornamental fish and aquacultural production. Ironically, in addition to being the taxonomic group with the most transgenic species, aquatic organisms are also the most likely group to present environmental concerns if accidentally released into the environment. Unlike most other agricultural species, fish are both difficult to contain and highly mobile, and they can easily become feral and invade native ecosystems (NRC 2002).

Transgenic Fish Defined

Transgenic fish are those that carry and transmit one or more copies of a recombinant DNA sequence (i.e., a sequence produced in a laboratory using in vitro recombinant DNA techniques). They are defined by the technology that is used to create and transfer the DNA sequence, not the source species of the donor DNA. Therefore, fish engineered with recombinant DNA derived entirely from fish are considered transgenic.

The recombinant DNA sequence, or construct, is usually comprised of several different regions including a start signal or "promoter," the coding region for the target protein, and a stop signal or "terminator." The construct is usually introduced into the animal's genome through microinjection of the recombinant DNA fragment into fertilized eggs or early embryos.

Inducing transgenesis is a relatively inefficient process. Only about one out of every 100 eggs microinjected will stably incorporate the recombinant DNA sequence into its genome and subsequently transmit the transgene to its progeny. The growth hormone gene has been the most popular target gene for transgenesis, which is not surprising considering the potential cost savings in feed for such a product. At least 14 species of fish have been genetically modified for enhanced growth, and although they almost always grow faster than nontransgenic controls, they do not necessarily grow to a larger mature size.

There are, however, some startling examples of gigantism (Nam et al. 2001). Several studies have shown that growth-enhanced transgenic fish have improved feed-conversion efficiency (Cook et al. 2000), resulting in economic and potential environmental benefits such as reduced feed waste and effluent from fish farms. Currently, no transgenic animal has been approved for food production in the United States, although that may change. A company called Aqua Bounty is currently awaiting regulatory review of its fish by the U.S. Food and Drug Administration (FDA).

Growth-Enhanced Salmon Proposed

Atlantic salmon remains the most important farmed food fish in global trade. Salmon is a carnivorous fish, and aquaculturalists have been working to improve feed-conversion rates and efficiencies through selective breeding, and the inclusion of plant-based protein (soy, rapeseed oil and corn gluten) in feed formulations. As a consequence, feed input per fish has decreased to 44% of 1972 levels; likewise, current diets contain approximately half the content of fishmeal that they once did (Aerni 2004).

The first transgenic food animal to be submitted for regulatory approval in the United States was transgenic Atlantic salmon carrying a chinook salmon growth-hormone gene controlled by a cold-activated promoter from a third species, the ocean pout. The mature weight of these fish remains the same as for other farmed salmon, but their early growth rate increases by 400% to 600%, with a concomitant 25% decrease in feed input and a shortened time to market (Du et al. 1992). Assuming a positive regulatory approval decision and consumer acceptance, the enhanced growth rate and feed efficiency of these transgenic salmon could increase salmon aquacultural productivity significantly, and would likely necessitate that salmon aquaculturists adopt the technology to remain competitive (Aerni 2004).

Risk Factors of Transgenic Fish

Release or Escape

The greatest science-based concerns associated with transgenic fish are those related to their inadvertent release or escape. Concerns range from interbreeding with native fish populations (Muir and Howard 2002) to ecosystem effects resulting from heightened competition for food and prey species. In principle, there is no difference between the types of concerns associated with the escape of genetically engineered fish and those related to the escape of fish that differ from native populations in some other way, such as a captively bred population (Lynch and O'Hely 2001). Ecological risk assessment requires an evaluation of the fitness of the transgenic fish relative to nontransgenic fish in the receiving population, to determine the probability that the transgene will spread into the native population. Fitness is defined as the genetic contribution by an individual's descendants to future generations of a population. It can be reduced to six net fitness components: juvenile viability, adult viability, age at sexual maturity, female fecundity (number of eggs), male fertility and mating success (Muir and Howard 2001).

The importance of accurately estimating each of the components of net fitness is demonstrated by the hazard exemplified by the "Trojan gene hypothesis." In this specific situation, the transgene confers enhanced mating success, but individuals possessing the transgene produce offspring with reduced juvenile viability. Depending upon the relative magnitude of the effects, an outcome associated with this particular set of circumstances can be the demographic destabilisation and ultimate extinction of the native population (Muir and Howard 1999; Hedrick 2001). It is therefore important to evaluate each species and transgene combination on a case-by-case basis to estimate the components of net fitness relative to nontransgenic fish in the receiving population (Muir and Howard 2004).

Environmental Factors

In addition to interbreeding, it is also important to consider the potential impact that environmental factors may have on the survival of transgenic and nontransgenic populations (i.e., genotype-by-environment interactions).

A recent study of growth-enhanced transgenic and nontransgenic salmon found that transgenic salmon did not affect the growth of nontransgenic cohorts when food availability was high (daily feed ration equivalent to 7.5% of total fish biomass). However, the survival of both transgenic and nontransgenic cohorts was deleteriously affected when feed resources were limited to 0.75% of total fish biomass.

The fast-growing transgenic salmon were found to dominate feed acquisition and exhibit strong agonistic and cannibalistic behaviour toward their cohorts when there were inadequate feed resources (Devlin et al. 2004).

Hunger and increased growth rates have been previously associated with agonistic behaviour in nontransgenic salmonids, although in this experiment, unmodified populations receiving the reduced feed ration did not display such behaviour.

The presence of transgenic fish will not a priori result in catastrophic results for native populations.

If transgenic fish are ill suited to an environment or are physically unable to survive outside of containment, then they may pose little risk to native ecosystems. It is important to realise that neither the risks nor the benefits of transgenic fish are certain or universal.

Both may vary according to a number of factors including the introduced gene, host species, containment strategy, species mobility, ability to become feral, relative fitness of the transgenic fish, receiving ecosystem, genotype-by-environmental interactions, and the stability of the receiving community. Regulators need to apply a scientifically sound, risk-based framework to assess the ecological risks involved with each transgene, species and receiving ecosystem combination on a case-by-case basis.

Simple Selective Breeding Programmes

Even though selective breeding programmes can be designed to improve all sorts of production phenotypes, the single most important phenotype is growth rate.

Improving growth rate will decrease the time it takes to grow a fish to market, which means a farmer can produce more crops in a given time period, which means he will make more money.

It also means yields will increase. This will increase production efficiency, increase food production, and also increase a farmer's income. As an added bonus, increasing growth rate can improve other production phenotypes via indirect selection.

Some studies have shown that faster-growing fish also have more efficient feed conversions and seem to be more disease resistant.

Selection can be used to improve other quantitative phenotypes, if improving them will improve production efficiencies or profits. In most cases, as important as these phenotypes are, they are not as important as growth rate.

Occasionally, a breeding programme other than selection is needed to solve the most important goal. For example: in tilapia farming, controlling reproduction in grow-out ponds is the most important goal, and this is accomplished by interspecific hybridization and/or sex reversal; in grass carp farming in the U.S., producing sterile fish is the most important goal, and this is achieved by chromosomal manipulation.

The object of this chapter is to outline simple, relatively inexpensive selective breeding programmes that can be conducted by farmers on farms with about 2 ha of ponds and to provide examples of the types of data that must be collected, as well as data tables that can be used to record these data.

This chapter will not discuss selective breeding programmes that use tandem selection; a selection index; or those that combine selection with crossbreeding, inbreeding, or some facet of biotechnology such as chromosomal manipulation.

When describing these simple selective breeding programmes, information will be provided on the number of ponds that will be needed in order to grow the fish that will be evaluated. This number will not include the number of ponds that will be needed to hold or to spawn brood stock. Additionally, this will not discuss ancillary facilities that are needed, such as holding tanks, hatchery buildings, counting tables, etc.

The main goal for the selective breeding programmes that are outlined here, will be to improve growth rate by selecting for length. As was described earlier, growth can be improved by selecting for either length or weight. Because it is far easier to accurately measure hundreds of fish than it is to accurately weigh them, and also because most farmers do not have access to accurate balances, most farmers should select for length.

Examples of selective breeding programmes that are designed to improve two phenotypes will also be outlined. In these examples, selection to improve growth rate is still the primary goal, but a second phenotype is added.

The selective breeding programmes outlined herein are designed to be conducted by a farmer who is creating genetically improved stock for his farm. If a farmer wants to embark on a breeding programme so that he can sell genetically improved fingerlings to support a local or regional industry, the size and costs of the projects that are outlined must be increased: more ponds must be built; more cohorts must be created; and more select brood fish must be produced. In fact, if this is a farmer's goal, it is likely that he will convert his entire farm to this enterprise, and he will no longer produce food fish.

One of the things a farmer must do if he is to run a successful selective breeding programme is to determine how many select brood fish must be saved. If too few are saved, a farmer will be unable to produce enough select fish for future grow-out. If this happens, he will have to spawn unselected fish, which will negate much of his efforts.

An additional problem is that if a farmer saves too few fish, inbreeding can build to levels which will result in inbreeding depression. If this occurs, much of the improvements created by selective breeding would be used simply to counteract inbreeding depression.

A farmer should spawn at least 25 males and 25 females every generation to minimise inbreeding depression. This number is not carved in stone, but it can be used as a general rule. This means that a farmer should save a minimum of 100–200 brood fish. The reason why a farmer needs to save this many select brood fish is some will die before they can be spawned, and spawning success is seldom 100%. These guidelines also apply for a control population.

A second factor that determines how many brood fish a farmer needs is the size of his operation; that is, how many fingerlings are needed for the grow-out ponds. If a farmer saves too few brood fish, he will not be able to stock his grow-out ponds. Saving enough select brood fish can be a problem on large fish farms, but it is seldom a problem on medium-sized (2 ha) fish farms, especially if a farmer saves at least 100–200 select brood fish.

The selective breeding programmes described herein are bare-boned skeletal outlines that are presented to demonstrate that relatively simple and relatively inexpensive programmes can be conducted by farmers and that, if the programmes are conducted properly, they can be integrated into everyday farming practices. The programmes outlined herein are not absolute. They can be and should be modified in order to customize the programme for a particular species, for a farmer's income, and for his farm.

The breeding programmes that are outlined here, are for species that do not exhibit sexual dimorphism. This means a single cut-off value can be used during selection.

If the species that a farmer grows exhibits sexual dimorphism a slight modification of the programmes that are outlined will be necessary. Sexual dimorphism can increase the effort needed to conduct selection, because once it occurs, the fish must be sexed and selection must occur independently in each sex. The age at which sexual dimorphism occurs will determine if a single cut-off value can be used or if separate cut-off values will be needed for males and females. In some species, sexual dimorphism occurs when the fish are small and separate cut-off values are needed at both the fingerling and food fish phases of selection. If sexual dimorphism does not occur until after the fingerling stage, a single cut-off value can be used at the fingerling stage, while separate ones will be needed when food fish are harvested. If sexual dimorphism occurs after the age at which food fish are harvested, a single cut-off value can be used during both phases of selection.

The selective breeding programmes outlined here, do not describe the creation and use of a control population to assess the results of selection.

Individual Selection

If possible, individual selection should be used to improve growth rate. Individual selection is easier and less expensive than family selection, because it can be done in only one or two ponds and fewer fish need to measured.

Selection for Growth Rate

Growth rate is the most important production phenotype. The following programmes outline how individual selection can be used to improve only this trait. When conducting individual selection, if a farmer can synchronize spawning, selection will be relatively simple. If the species spawns asynchronously, care must be taken to ensure that age-related size differences do not confound and obscure genetically-produced size differences. If the selective breeding programme is not modified to accommodate asynchronous spawning behaviour it will be impossible to identify fish which are genetically superior from those which are environmentally superior (older).

Synchronous spawning: If a farmer can synchronize spawning and can produce at least 25 families on a single day (or at most, over a 48-hour period), he can conduct a simple, inexpensive selective breeding programme in only one or two ponds. In addition, an extra pond will be needed for the select brood fish, once they have been saved.

To initiate the selective breeding programme, the fish should be spawned using normal management techniques. If possible, egg masses should be collected and incubated. If the eggs are usually incubated by the females in the ponds, they should be closely monitored, and fry should be collected as soon as they hatch or begin to swim away from each female.

If possible, family size should be equalized before fish are stocked. This will prevent one family from skewing the results of selection; additionally, it will help minimise inbreeding. Because of this, families should be isolated until they are equalized. In addition, if families are isolated until they are stocked, the complete mortality of one or more families will be noticeable and can be recorded.

If the fish are traditionally raised in a two-phase process (phase one is stocking fry and raising them to fingerlings; phase two is stocking fingerlings and raising them to food fish), selection can be done when fingerlings are harvested and when fish are harvested for market; if this is the case, selection will require two ponds. If production is only a single-phase process (no fingerling phase; fry are stocked and raised to food fish), selection will occur when fish are harvested for market, and selection will require only one pond.

If fish are raised in a two-phase process, fry should be stocked in a single pond and fingerlings should be produced using normal production techniques. Just prior to harvest, a random sample of 100–200 fingerlings should be measured to the nearest millimetre in order to determine the phenotypic value that corresponds to the desired cut-off percentage. At fingerling harvest, the top 35–50% should be saved and stocked in the food fish production pond. The culled fingerlings can be grown for food or sold. If they are grown for food, no fish from this population should be saved and spawned.

The select fingerlings should be raised using normal production management. Prior to harvest, a random sample of 100–200 fish should be measured to the nearest millimetre to determine the phenotypic value that corresponds to the desired cut-off percentage. At harvest, the top 10–20% of the fish should be saved, and these fish will become the select brood fish. Culled fish can be eaten or sold.

If fish are raised in a single-phase process, a random sample of 100–200 fish should be obtained to create the F_1 control brood fish when fish are harvested. The population of F_1 control brood fish should be created prior to selection. If fish are raised in a two-phase process and if selection will occur at both the fingerling and food fish harvests, the F_1 control brood fish need to be saved at the end of the fingerling phase. If the F_1 control brood fish are saved just prior to the second act of selection during a two-phase selection process, the control will not be a true control population, since some selection will have occurred prior to its creation. If fish are raised in a single-phase process, fry should be stocked in a single pond, and fish should be produced using normal production techniques.

The cut-off percentages were mentioned as ranges, not as specific values (35–50% for fingerlings and 10–20% for food fish). This is because the exact cut-off values are not that critical. The intensities of selection that are used are individual decisions. A farmer can increase the rate of gain by using higher cut-off percentages (saving a smaller percentage). But rate of gain must be balanced against inbreeding-related problems and with the ability to produce enough fish in the next generation. Most selective breeding programmes conducted on medium-sized farms will encounter few problems and will also be able to achieve desired results if cut-off values in these ranges are used.

The pond(s) used in this project should be 0.04–0.4 ha. Pond size is determined by the stocking rate that will be used to grow the fish and by the intensity of selection. Once a farmer decides what his cut-off percentages will be and how many select brood fish will be saved, he can determine how big the ponds must be. This will tell him the percent of the farm that will be devoted to the breeding programme.

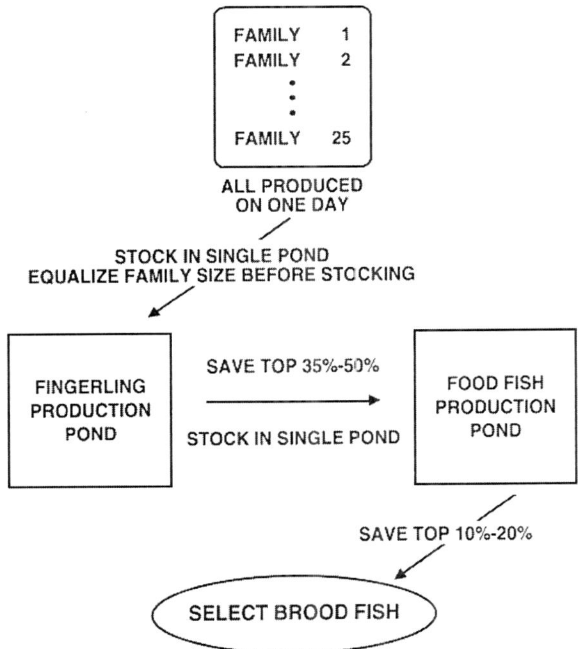

Figure: *Schematic diagram of a simple and inexpensive selective breeding programme to improve growth rate by selection at two ages-at fingerling harvest and at food fish harvest. This breeding programme can be conducted in two ponds. Fish that are culled at the fingerling stage can be sold or grown for food. Fish that are culled at harvest can be eaten; sold as food; or some can be retained and used as brood fish to produce the production fish (fish raised and sold as food), if the select fish cannot produce enough offspring for both the selective breeding programme and for the production ponds.*

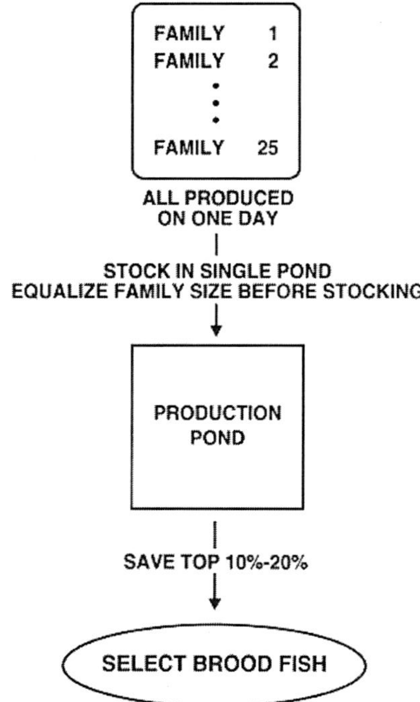

Figure: *Schematic diagram of the least expensive selective breeding programme that can be designed to improve growth rate. In this breeding programme, fry are stocked in a pond, and the fish are not harvested until they are ready for market. Selection occurs when the pond is drained and the fish are harvested. This breeding programme can be done in one pond.*

If ponds of approximately the correct size already exist, a farmer does not have to destroy them and custom-build new ponds in order to conduct a selective breeding programme. A farmer who uses existing ponds should adjust stocking rates and/or cut-off values to achieve the desired results. The information generated in Table below can also be used to determine how many fish are needed from each family. For example, if a farmer needs to stock 8,892 fry and if he has produced 25 families, he needs 355.7 fry/family (8,892/25). Because you cannot have 0.7 fry, he needs 356 fry/family. Because the number is rounded up, the farmer will stock 8,900 fry. If he maintains the stocking rate at 200,000 fry/ha, the size of his fingerling pond would change only by 0.001 ha to accommodate the 8 additional fry, so a larger pond would not be needed (pond size had already been rounded to the nearest thousandth). The rounding up of family size also means that the initial culling will produce an additional 2 select fingerlings. This will have little effect on stocking rate in the food fish pond, nor will it require a larger food fish pond. If the farmer does not want to alter stocking rate one iota from his planned rate, he can randomly cull 2 select fingerlings once that population has been created.

Brood fish that are saved should be stocked in a select brood stock pond. No other fish should be stocked with the select brood fish. When mature, these fish will be spawned to produce the F_1 select generation. No selection for secondary sexual traits and no selection for other criteria should be done with the brood fish prior to the mating season. The only goal of this selective breeding programme was to improve growth rate by selecting for length, and the select brood fish were selected for that phenotype and should be selected only for that phenotype; they should not be selected for any other reason.

Table: *Procedure that can be Used to Determine the Size of the Ponds that are Needed in a Selective Breeding Programme.*

Goal:	To have 200 select brood fish
Given:	Cut-off at food fish harvest: select top 10% Cut-off at fingerling harvest: select top 50% Fry-fingerling mortality: 50% Fingerling-food fish mortality: 10% Stocking density in fingerling pond: 200,000/ha Stocking density in food fish pond: 7,000/ha
Step 1.	How many food fish must be harvested to produce 200 select fingerlings, if you save the top 10%?
	Number of food fish harvested = number saved/percent saved = 200/0.1 = 2,000.
Step 2.	How many select fingerlings should be stocked in the food fish pond, if mortality is 10%?If mortality is 10%, survival is 90%:
	Number of fingerlings stocked = number harvested/survival rate = 2,000/0.9 = 2,223.
Step 3.	How many fingerlings must be harvested from the fingerling pond to produce 2,223 select fingerlings, if you save the top 50%?
	Number of fingerlings harvested = number saved/percent saved = 2,223/0.5 = 4,446.
Step 4.	How many fry must be stocked in the fingerling pond to produce 4,446 fingerlings, if mortality is 50%? If mortality is 50%, survival is 50%:
	Number of fry stocked = number harvested/percent survival = 4,446/0.5 = 8,892.
Step 5.	How big should the fingerling pond be, if it is stocked at 200,000/ha and if you will stock8,892 fry?
	Size of fingerling pond = number of fry that will be stocked/stocking rate =8,892/200,000 = 0.0446 which is rounded to 0.045 ha.
Step 6.	How big should the food fish pond be, if it is stocked at 7,000/ha and if you will stock2,223 fingerlings?
	Size of food fish pond = number of fingerlings that will be stocked/stocking rate =2,223/7,000 = 0.3175 ha which is rounded to 0.318 ha.

The select brood fish should be allowed to mate among themselves in a random manner or, if they will be paired in pens or manually stripped, they should be randomly paired.

If a farmer has enough select brood fish, he can use these fish to produce both the F_1 select generation and the production fish (fish that the farmer will stock and grow for market in his production ponds).

If the select brood fish cannot produce a sufficient number of offspring for both purposes, their fry should first be used to produce the F_1 select generation; any surplus fish can be grown for food. If the select brood fish cannot produce enough fingerlings for both purposes, the farmer should use other brood fish (culls from the harvest selection) to produce fry for the production ponds.

The second, third, etc. generation of selection can proceed as described above. If the select brood fish cannot produce enough fish for both the selective breeding programme and for the production ponds, genetic improvements that are being achieved in the select population can be transferred to the production fish beginning with the second generation of selection, if the farmer uses the culls from the breeding programme as brood fish to produce fry for his production ponds.

Even though these brood fish were culled, they came from the selective breeding programme, and their parents were select brood fish. This will enable a farmer to transfer the genetic gain to the farmed fish, but it will be with a one-generation delay.

Additionally, when a farmer replaces one generation's select brood fish with their successors, he can transfer the previous generation's select brood fish to the production brood fish population. This will also enable the farmer to transfer genetic gain to the production population.

If selection will occur only when food fish are harvested, a farmer will not be able to transfer the genetic improvement to the production population as quickly, unless the select fish can produce enough offspring for both the breeding programme and for the production ponds. If the select brood fish can produce only enough offspring for the breeding programme, a farmer must use unselected brood fish to produce fry for the production ponds after the first generation of selection.

After the second generation of selection, he can either use F_1 select brood fish or fish that were culled when he created F_2 generation brood fish. This means a farmer will realise no genetic improvement in his production fish for one generation, but thereafter, he will be able to transfer it with a one-generation delay.

Asynchronous spawning: If a farmer cannot spawn his fish synchronously, he should divide the population into age cohorts and select for growth rate independently within each cohort. To initiate this breeding programme, a farmer should spawn the fish using normal management techniques. As before, it is better if egg masses can be collected and incubated. If the eggs are usually incubated by the mother in the ponds, females should be closely monitored, and fry should be collected from each female as soon as they hatch or begin to swim away.

Egg masses or newly hatched fry should be grouped into daily (24-hour) age cohorts. If it is not possible to collect enough families within a 24-hour period, the time interval for each cohort can be stretched to 48 hours.

Each cohort should be composed of at least 5 families, and there should be at least five cohorts. These numbers are not carved in stone. A farmer does not have to discard a cohort simply because it is made up of only four families. There are two basic premises behind this work plan: the first is that a cohort should be composed of several families; the second is that at least 50 parents (25 males and 25 females) should produce offspring for the breeding programme.

As before, if this species is usually produced by a two-phase production process, selection for increased length will occur when fingerlings are harvested and when food fish are harvested. If production is usually a single phase with no fingerling phase, selection will occur only when fish are harvested for food. Individual selection will be used to select the best fish from each cohort.

This breeding programme will require five to ten 0.04-ha ponds. The number of ponds depends on the size of the project (how many cohorts will be produced) and whether production is a two-phase or a one-phase process. The exact size of the ponds can be determined as was described previously. The exact size is not that important, but all ponds should be similarly sized. The reason you want small ponds is because a large number can be squeezed into a small space. Additionally, it is cheaper to build a 0.04-ha pond than a 0.1-ha pond.

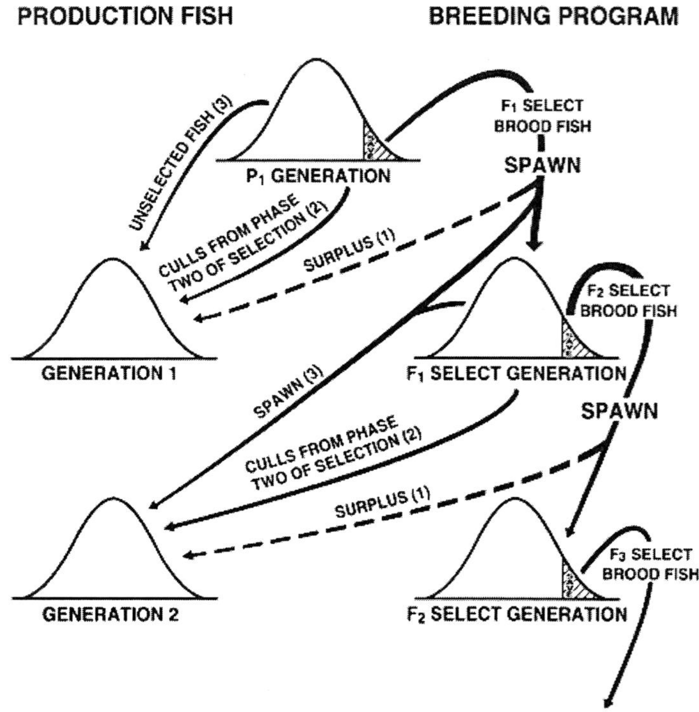

Figure: *Schematic diagram illustrating the ways genetic gain can be transferred from the selective breeding programme to the production fish that a farmer grows for market. If the select brood fish can produce enough for both populations, the transfer will be immediate, and the means of the two populations will be the same (path 1). If selection is a two-phase process, the culls from the second phase can be used as brood fish to produce the production fish (path 2). If this approach is used, some of the genetic gain will be transferred immediately, but the production mean will always lag behind the select mean. If selection is done only at harvest, unselected fish will have to be used to produce the first generation of production fish, and either the culls from the breeding programme or the previous generation's select brood fish can be used thereafter (path 3). This approach will transfer the gains with a one-generation delay if the previous generation's select brood fish are used; if culls from the breeding programme are used to produce the production fish, the production mean will be slightly better than the mean of the previous generation of select fish. All assumptions about mean values were generated using the premise that there is no environmental influence on the phenotype and the trait has a large heritability.*

If selection is a two-phase process, select fingerlings could be restocked in the ponds where they were produced, if the ponds can be drained and refilled in a single day. This would reduce the number of ponds needed for the programme by one-half. The only prerequisites are: all fish must be harvested; the farmer must have holding facilities where he can safely hold the select fingerlings until they are restocked.

Farmers who cannot afford to build the ponds could conduct this breeding programme in large hapas (20–40 m^2) that are placed in one or more ponds. Even though hapas are less expensive than ponds, it is better to use ponds, because if the fish are grown in ponds they should be selected on the basis of growth in ponds, not on the basis of growth in hapas.

Each cohort should be stocked in a single pond; stocking rates for each cohort should be identical, but it is not critical, since selection will occur independently in each cohort. Even though it is not necessary to have identical stocking rates in all ponds, it is advisable to have similar stocking rates, or selection could select for slightly different genes in the different cohorts.

The best way to create the population of fish that will be stocked in each pond is to choose an equal number of fish from each family within a cohort. The fish chosen from each family must be randomly selected; they cannot be chosen because they are the largest, etc. Therefore, families must be isolated until family size is equalized. If the families within a cohort are mixed before the proper number is chosen to create the desired stocking rate, the largest family will be over-represented and the smallest will be under-represented.

When the fingerlings are harvested, the top 35–50% from each cohort (pond) should be stocked in a single grow-out pond. To determine the fingerling cut-off value for each cohort, a sample of 100–200 fingerlings from each cohort should be measured to the nearest millimetre, and the phenotypic value that corresponds to the desired cut-off percentage should be determined. The cut-off value must be determined independently for each cohort, because they are managed as temporary, separate sub-populations during the selection process, and selection will occur independently in each cohort.

As before, the intensity of selection is not that critical, but it must be the same for each cohort. Fish from different fingerling ponds (cohorts) should not be mixed when they are harvested and measured. After selection, the select fingerlings from each cohort should be stocked in a separate food fish ponds. Fish from different cohorts should not be mixed.

As was the case when stocking the fingerling ponds, the stocking rates for the food fish ponds should be the same or at least similar. The select fingerlings in each cohort should be grown using normal production techniques.

Just prior to harvest, a random sample of 100–200 fish from each cohort should be measured to determine the cut-off value for each cohort. The top 10–20% from each pond (cohort) should be saved to form the population of select brood fish. Again, the intensity of selection is not that critical, but it must be the same for all cohorts.

Once the select brood fish have been chosen from each cohort, they can be mixed and stocked into one or two select brood stock ponds. These ponds should contain no other brood fish. As before, a major goal of the selection process is to have at least 100–200 select brood fish.

The brood fish should be managed and used to produce fry for the breeding programme and for the production ponds, as was described in the previous sub-section.

If this selective breeding programme were going to be conducted by a scientist at a research station, a major aspect of his experimental plan would be to make all aspects of management identical for all cohorts. But a farmer does not have to worry about this.

It would be nice if all cohorts were managed identically during each phase of selection, but because each cohort is stocked in a single pond and because selection will occur independently within each cohort, minor management differences among the cohorts will not affect selection.

For example, if one food fish pond is stocked with 5,000 fingerlings/ha and the others are stocked with 4,000 fingerlings/ha, the different stocking rates could affect the average growth rate of the cohorts, but since selection will occur independently in each pond (cohort), it probably will not affect the selection process.

However, large differences in management among the ponds could affect the outcome. For example, if one food fish pond is stocked with 15,000 fingerlings/ha and if cattle manure is used as the sole source of nutrients while the others are stocked with 4,000 fingerlings/ha and if fish are fed rice bran, selection may select for different genes in the differently managed cohorts.

Selection for Growth Rate and Another Phenotype

The breeding programmes that have been outlined can be expanded to include another phenotype. If a farmer wants to improve phenotypes other than growth rate, he can easily add traits such as body conformation and/or harvestability. Farmers should use independent culling or modified independent culling to simultaneously improve two phenotypes. Aquaculturists at fingerling production centres should also use independent culling; however, they could use a selection index if they are technically sophisticated and have the expertise and labour needed to conduct this programme.

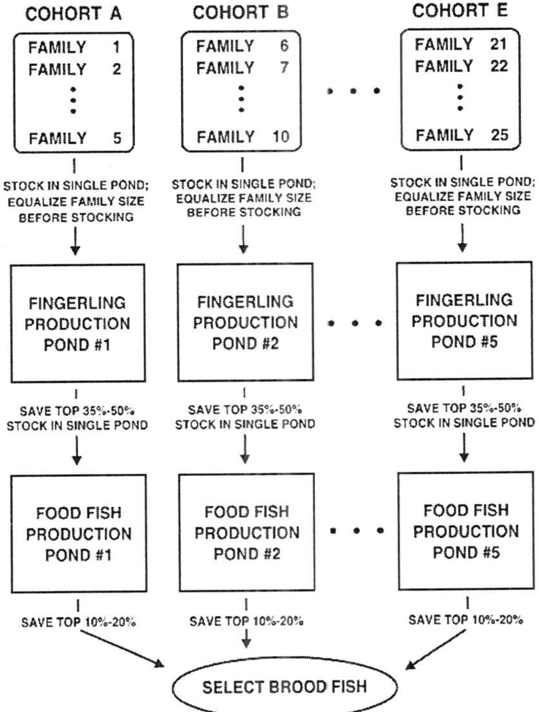

Figure: *Schematic diagram of a selective breeding programme that uses individual selection to improve growth rate when a farmer cannot synchronize spawning. This selective breeding programme divides the population into age cohorts, and selection occurs independently within each cohort. In the programme outlined in this figure, there are five cohorts (A-E), and there are five families within each cohort. If there is no fingerling phase, step one can be eliminated, and the selective breeding programme can be conducted in five ponds. If the number of families can be increased in each cohort, fewer cohorts will be needed, which will decrease the number of ponds that are required.*

When independent culling is used, a farmer needs to determine the overall intensity of selection in order to calculate the cut-off percentage for each trait. A farmer should try to save 10–20% of the population at the food fish phase of selection. The phenotypic values that correspond to the desired cut-off percentages should be determined as was described previously.

Body conformation is an important phenotype, and improving this trait can increase yields. A deeper-bodied fish or a thicker-bodied fish will carry more muscle on its frame, which means each fish weighs more per centimetre body length than a normal, streamlined fish. Selecting for weight might improve body conformation, but this type of selection does not guarantee that body conformation will be improved. Selecting for weight simply improves average weight, and heavier fish might be fish that are longer, that have larger heads, etc. Improving body conformation has been a continuing goal of many cattle, swine, sheep, and poultry breeding programmes, because animals with better body conformations produce more meat.

One way to improve body conformation is to select for both length and body depth at the anterior margin of the dorsal fin (at the first dorsal fin spine). Since the primary goal of the breeding programmes that are outlined herein, is to select for length as a way of increasing growth rate, all a farmer has to do is add selection for body depth, and he will add body conformation as the second phenotype.

Alternatively, body conformation can be improved by selecting for a body length:body depth ratio. This has been done with common carp and it succeeded. However, even though body conformation was improved, average weight did not increase. This occurred because fish were selected only for the body length:body depth ratio. Consequently, small, but deep bodied fish could have become select brood fish. To improve both growth rate and body conformation, both traits should undergo selection.

If selection will occur at the end of both the fingerling and food fish phases of production, the initial act of selection (fingerling phase) can be for length only, as was described earlier. The rate of progress will be slower for body conformation (body depth) than for growth rate (length) if selection for body depth is conducted only on food fish, but if independent culling is conducted twice, a farmer may end up with only a few select brood fish.

When fish are raised in ponds and harvested by seining, harvestability is a phenotype that a farmer might wish to improve. Anyone who has seined a pond knows all too well that fish are experts at escaping a seine.

Most farmers never consider the costs of harvesting when they determine annual production budgets, but it can be significant in terms of labour and equipment. Hard-to-capture fish can become stressed and killed as a result of repeated seinings. Finally, fish that are not captured cannot be sold or eaten.

If a farmer wants to produce fish that are easier to capture, he can add this phenotype to his selective breeding programme and select for both growth rate (length) and harvestability by using independent culling. If a farmer decides to select for harvestability, he needs to define it as "fish that are captured during the first seine haul."

If this is done, independent culling will be a two-step process: The initial step will be to save only those fish that are harvested during the first seining; all other fish will be culled. The second step will be to select for length in the fish that were saved (seined).

It may be that so few fish are harvested during the first seining that a farmer cannot select for length efficiently or meaningfully. If this is the case, the goal should be modified so that fish that are captured during the first two seinings are saved.

The ability to escape a seine is not only a detrimental phenotype in that it increases production costs, it is detrimental because it can produce slower-growing fish, if a farmer conducts selection for growth rate improperly. If a farmer cultures fish using multiple batch production, selection for growth rate can be done only during the first harvest after the pond is filled.

Thereafter, size and age are confounded, especially if fingerlings are stocked to replace fish that are harvested or if fish can reproduce in the pond.

If a farmer chooses to select for growth rate and for a second phenotype, growth rate will be improved more slowly than it would be if it were the only phenotype under selection. A farmer could select for growth rate, harvestability, and body depth (or any other phenotype he wants to improve) by using independent culling, but the rates of improvement for all three traits would be small.

Family Selection

In general, family selection is used when heritability is small and/ or when there are uncontrollable sources of environmental variance which obscure genetic differences and which make individual selection ineffective.

Within-family selection is usually used when there is a large environmental source of variance that has a major influence on phenotypic variance at the family level. Chief among these factors are spawning date and age and size of the mother.

Between-family selection is usually used when most of the phenotypic variance is due to environmental sources of variance, and they are felt at the individual rather than at the family level. When this occurs, an individual's phenotypic value does not accurately reflect its breeding value; consequently, individual selection will be ineffective and between-family selection must be used. By comparing family means, a farmer can neutralize much of the environmental component of phenotypic variance, and the family means can be used to assess the average breeding value of every fish in each family.

Between-family selection is also used when animals must be killed before their phenotype can be measured. For example, between-family selection is usually used to improve carcass traits, because animals have to be slaughtered in order to be measured.

Within-Family Selection

The simplest type of family selection that can be used is within-family selection. This is because each family can be considered a separate sub-population, and selection will occur independently within each family, in a manner similar to that described for individual selection when the population was divided into age cohorts.

In this case, each family can be considered a cohort. Unless families can be given permanent family marks, families will have to be cultured in individual ponds or hapas.

The size of the ponds that are needed in this type of breeding programme will depend on the fecundity of the species. With some species, pond size could be as small as 100 m^2. For example, despite their reputation for being incredibly prolific, tilapia produce relatively small families. In general, family size ranges from 50–1,500, depending on the female's size. If the desired stocking rate is 5,000/ha, a family of 50 would have to be stocked in a 100-m^2 pond to achieve that stocking rate. If family size is 50–100 fish, it might be more efficient to raise the families in 10-to 20-m^2 hapas that are suspended in a 0.1- to 0.2-ha pond.

Twenty-five to 50 families should be produced and entered into the selective breeding programme. This means that 25–50 ponds must be constructed for the project.

If the ponds can be drained and filled in a day and if selection is a two-phase process, the number of ponds that are needed can be reduced by one-half if the farmer has holding facilities. When the ponds are refilled, the fingerlings from each family can be restocked into the pond where they were produced.

Twenty-five families is the minimum number that should be entered into this selective breeding programme, because you want 25 males and 25 females to produce offspring for reasons that were described earlier. If a farmer wants to use the minimum number of families, he should produce 27–35 families, because mortalities in some families may reduce family size below that needed to stock a pond with the desired number of fry.

Fish should be spawned and families should be produced as was described for individual selection. Families must be isolated, because selection will be at the family level.

Because each family will be treated as a temporary sub-population and because selection for length will occur independently within each family, minor differences in management among the ponds (families) will not affect the selective breeding programme. This means that a family does not have to be discarded because a farmer cannot stock it at the desired rate. As was the case for individual selection when the population was divided into age cohorts, minor differences in management among the ponds are of little consequence, but a farmer should try to manage all ponds similarly.

At harvest, a sample of 30–100 fish from each family (or every fish if family size is small) should be measured to the nearest millimetre to determine where the cut-off should be placed. In this case, the cut-off value is expressed as the largest 5 or 10 or 20, etc. fish. A farmer should save the top 5–10 males and the top 5–10 females from each family. If no sexual dimorphism exists at harvest, a farmer can simply save the top 10–20 fish from each family.

This type of selective breeding programme obviously requires more effort than individual selection. If 25 families are raised and if a farmer measures 30–100 fish from each family, he will measure 750–2,500 fish to determine cut-off values, as opposed to the 100–1,000 fish that were needed in the programmes that were outlined for individual selection.

Furthermore, this type of selective breeding programme could stress the fish.

A farmer has to measure every fish in each family twice-once to establish the cut-off value and once to determine which fish should be saved.

Once the select brood fish from each family are saved, they can be mated using either of two techniques: The first and easiest is to stock the fish in a single pond and spawn them in a random manner.

If rotational mating is used, either each family must be given a unique mark and stocked communally until the next breeding season when they will be isolated once again for spawning, or each family must be stocked in a separate pond.

Rotational mating is very expensive in terms of facilities and labour, and it greatly increases the costs of the breeding programme. For these reasons, the first approach is recommended for most farmers.

The genetic gain should be transferred to the production fish as was described earlier. If possible the select brood fish should be spawned to produce both the F_1 select generation and fry for the production ponds.

If a farmer wants to improve two phenotypes by within-family selection, he can add a second phenotype such as body depth or catchability as was described earlier for individual selection.

A farmer must be judicious in his use of within-family selection to improve two or three phenotypes. If family size is small, only one or two fish from each family may be able to meet or exceed all cut-off values. If this occurs, a farmer must either greatly relax his cut-off values or evaluate 100–200 families, which might not be possible on a 2-ha farm.

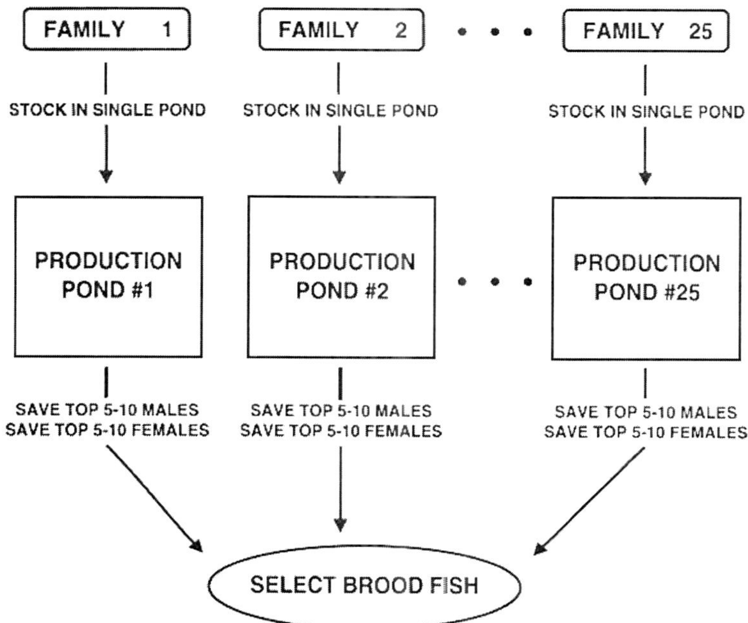

Figure: *Schematic diagram of within-family selection to improve growth rate. A minimum of 25 families should be evaluated, and each family can be stocked in a single pond.*

Between-Family Selection

Between-family selection is more expensive than individual selection and within-family selection, because it requires more ponds. Because whole families will be saved or culled, a farmer should enter 25–50 families. It is better if a farmer can evaluate 50 families, but the costs will probably be prohibitive for most farmers.

Because family means will be compared, a farmer must raise each family in at least three ponds, and the assignment of the families to the ponds must be random; consequently, a farmer will need 75–150 ponds for this programme.

When fish are measured to determine which families will be saved and which will be culled, the means from the three ponds are averaged, and it is the overall mean from the three ponds which is used as each family's mean.

Families must be raised in replicated ponds because this is the only way to isolate pond-related size differences from gene-related size differences. If each family were stocked in only one pond, the largest family might be the largest simply because the pond had the best algal bloom.

Because this type of selection requires the use of so many replicated ponds, it might be cost-prohibitive for farmers who only want to produce genetically improved fish for their own use. For example, if a farmer evaluates 50 families, he will need 75 ponds. If each pond is 0.01 ha, he will use 37.5% of his 2-ha farm for the breeding project.

This unfortunate side-effect of family selection can be circumvented only if a farmer is able to give each family a unique mark. If he can, the fish can be stocked communally in one or two 0.1- to 0.25- ha ponds. Even if this is possible, each family must be raised in an individual tank or pond until it can be marked. At harvest, the fish must be separated into families once again, so the farmer must have adequate holding facilities.

Because whole families will be saved or culled, even if production is a two-phase process, selection needs to be conducted only when food fish are harvested. If the farmer wants to select the families twice, he can cull the worst 5–10 families during the fingerling phase of selection. This will make the grow-out phase less expensive.

If there is a large genetic correlation between fingerling size and food fish size, selection can occur at the fingerling phase instead of the food fish phase. This will reduce the costs of the breeding programme because it will reduce the number of ponds that are needed for the food fish phase, since only the select families will be grown to market size. This approach has been used with rainbow trout, and it improved harvest size by indirect selection.

At the end of the food fish phase of production, the top 5–10 families should be saved; these are the select brood fish. A farmer can save either the entire family or a random and equal number from each select family. As before, a farmer should save at least 100–200 select brood fish.

The one aspect of between-family selection for increased growth rate that can be inexpensive is obtaining data to determine which families should be saved. Because selection is based on family means, the fish from each pond can be batch weighed. If the number of fish that are weighed is known, the mean weight can be easily determined. This will enable the farmer to improve growth rate by selecting for weight rather than for length.

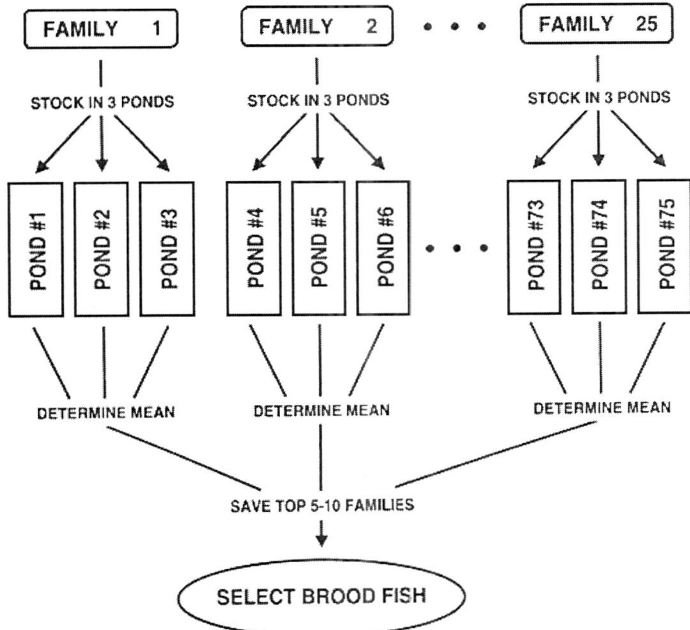

Figure: *Schematic diagram of between-family selection to improve growth rate. A minimum of 25 families should be evaluated. Because family means are compared, each family must be grown in at least three ponds, and the overall mean of the three ponds is the value that determines whether a family is saved or culled. If the fish can be ven family marks, they can be grown in a single pond, but the families must be segregated when they are measured and selected.*

In general, between-family selection precludes selection for two or more phenotypes, because only those families whose means meet or exceeded two cut-off values would be saved, and it is unlikely that the top five families for one phenotype would also be the top five for a second. A farmer could improve two phenotypes by using between-family selection to improve one phenotype and another type of selection to improve a second phenotype.

The select brood fish should be managed and spawned as was described for the within-family selective breeding programme. Again, because it is less expensive, most farmers should mix the select families and randomly spawn the select brood fish. However, if this is done, farmers should be aware of the fact that inbreeding will build to levels that will cause problems within a few generations. This occurs because between-family selection makes the breeding size of the population far smaller than the actual number of brood fish. This type of breeding programme is designed to save fish from only 5–10 families each generation, which means the breeding size of the P_1 generation is retroactively lowered to 10–20, and it is smaller thereafter.

A combinational approach that is often used is to combine between-family and within-family selection. If this combination is used to improve two phenotypes, the first step is to improve growth rate via between-family selection.

The second step is to use within-family selection to improve the second phenotype. Of course, between-family and within-family selection can be combined in a two-step selection programme to improve only growth rate. A logical approach, and one that has been successfully used with rainbow trout and coho salmon, is to use between-family selection at the fingerling phase of selection and to use within-family selection at harvest.

If the between-family selection and within-family selection combination is used, it will be expensive. It will combine all of the costs of between-family selection with some of the costs of within-family selection.

Record Keeping

Selective breeding programmes work only when farmers keep records. It is probably the least appreciated but most necessary aspect of any breeding programme. Without records, there is no way a cut-off value can be determined, so it is impossible to create a population of select brood fish. Without records, there is no way to determine if a selective breeding programme is succeeding. Without records, a farmer will not know which ponds contain select fish, which ponds contain select brood fish, which ponds contain control fish, and which ponds contain fish that are being grown for market.

Farming is one occupation where record keeping should be considered an integral part of everyday life. Farmers should gather information about spawning success, the stocking rates used, mean size at stocking, the amount of fertilizer used, the amount of feed used, percent survival, mean weight at harvest, yield, etc. If a farmer has this information, he knows what is going on and does not rely on guesses. He knows if yields decreased because of bad weather, and he knows how much they decreased. He knows if yields increased because he used a better quality fertilizer, and he knows how much they increased. If there are no records, he can only guess.

Farmers should keep records of these parameters for every pond. Ponds have individual personalities. Records will indicate which ponds are bad and which are good. This information can also be used to customize management for each pond.

Many farmers do not want to or are unable to keep accurate records. These farmers probably should not be encouraged to conduct a selective breeding programme. If a farmer will not gather the data that are needed to evaluate a programme, he probably will not conduct the programme properly. This could be detrimental on a regional basis because the farmer might tell his neighbours that selective breeding is a waste of time-he will not tell them that it did not work because he did not conduct it properly.

Extension agents could provide record keeping services for hard-working, reliable farmers who are unable to keep records. If selective breeding programmes are kept relatively simple, an extension agent could provide this service for a number of farmers. The only drawback to this approach is that these farmers will become totally dependent upon the extension agent and often will move from one phase of their programme to another only when the extension agent appears. If the extension agent is transferred to another region or retires, the selective breeding programmes could collapse if the new extension agent has other priorities.

What kind of data should a farmer take and what kinds of records should he keep if he is going to conduct a selective breeding programme to improve growth rate or other quantitative phenotypes? First and foremost, a farmer must be able to describe the phenotype he is going to improve via selection. This means he must be able to measure it accurately and quickly and without stressing the fish.

Farmers who take data routinely will be able to incorporate what is needed into everyday management. In fact, if a farmer already records the kind of information that was mentioned earlier in this section, he may be recording these data routinely, so there will be little extra work.

If a farmer is going to conduct a selective breeding programme to improve growth rate by individual selection, he must obtain and maintain data on: the number of fish that spawned, the number of families that were produced, and the dates the families were produced; the families that contributed offspring to the selective breeding programme; the number of fry from each family that were grown; when the fish were stocked; where the fish were stocked; when the fish were harvested; when the fish were measured; phenotypic values; the cut-off value; the spawning success of the select brood fish; the performance of the F_1 select generation, as well as that of the control population.

This process is then repeated for the second generation of selection, etc. In addition, a farmer should maintain data about normal, everyday management for every pond used in the selective breeding programme.

Examples of data tables that can be used to gather the information needed to conduct a selective breeding programme by individual selection are presented in Tables; data tables that can be used to record normal, everyday management will not be presented.

They are simply presented as examples. Any permutation may be used, as long as it is well organised and as long as the data are easily accessible.

Table Below are examples of data tables that can be used to record spawning information for each generation of individual selection. Table below is a data table that can be used to record data for a simple selective breeding programme, where the species can be spawned synchronously and where the fish will be grown in a single pond.

Table below is a data table that can be used to record data for a selective breeding programme where the species cannot be spawned synchronously, and where the fish will be raised and selected in cohorts. Both data tables provide data on the date each spawn was produced, the size of each family, the number of fish from each family that was used in the selective breeding programme, the pond where each family was stocked, and the date the pond was stocked.

Tables below are examples of data tables that can be used to record phenotypic values for individual selection. Table below is for a selective breeding programme that will select for only one phenotype, is for one that will incorporate two phenotypes.

The tables include information about the group of fish: when the group of fish was produced, when it was stocked, and when the fish were measured (so the age can be determined); the pond in which the population was stocked; the number of families that contributed fish to the population. Both tables are designed for species that do not exhibit sexual dimorphism.

If a species exhibits sexual dimorphism and if separate cut-off values will be used for each sex, the tables can be divided into male and female sections or separate tables can be kept for each sex.

Table: *Example of part of a data table that can be used to record spawning data for a selective breeding programme that will improve growth rate by individual selection; the fish will be stocked in a single pond. The table also includes data on the number of fish stocked per family, the pond in which they were stocked, and the stocking date.*

Dates: May 1 – May 15, , 1995 Pond Nos.: 1, 2, and 3 Stocked on : April 30, 1995					Species: Any fish species Generation: P_1 generation. Fish grown will be selected to become F_1 select brood fish.					
Date	Spawn	Pond	Weight egg mass (g)	Number of eggs	Number hatched	Number of fry	Used in breeding programme?	Number stocked	Pond stocked	Dated stocked
5/1	1	2	310	9.300	8.370	7,533	no	0	-	-
5/3	2	1	262	7,860	6,681	5,946	yes	150	8	5/8
5/3	3	1	138	4,140	3.643	3,169	yes	150	8	5/8
6/3	4	1	152	4,560	4.332	4,115	yes	150	8	5/8
6/3	5	1	147	4,410	3.176	2,889	yes	150	8	5/8
5/3	6	2	25	750	325	145	no	0	-	-
5/3	7	2	273	8,190	6,879	5,641	yes	150	8	5/8

Table: *Example of a data table that can be used to record spawning data for a selective breeding programme that will improve growth rate by individual selection; the population will be divided into age cohorts. The table also includes data on the number of families in the cohort, the number of fish stocked from each family, the pond in which the cohort was stocked, and the stocking date.*

Dates: May 5 – May 15, , 1995 Pond Nos.: 3, 4, and 5 Stocked on : May 1, 1995					Species: Any fish species Generation: P_1 generation. Cohort A. Fish grown will be selected to become F_1 select brood fish.				
Date	Spawn	Pond	Weight egg mass (g)	Number of eggs	Number hatched	Number of fry	Number stocked	Pond stocked	Dated stocked
5/5	1	3	413	12,390	11.151	9,924	200	12	5/10
5/5	2	3	352	10.560	8.973	8.257	200	12	5/10
5/5	3	4	238	7.140	5,640	4,906	200	12	5/10
5/5	4	4	252	7.560	6,350	5.715	200	12	5/10
5/5	5	4	136	4,080	3,876	3,565	200	12	5/10
5/5	6	5	251	7,530	5,647	5,251	200	12	5/10

Table: *Example of a data table that can be used to record length measurements at harvest. This data table is designed for a species that does not exhibit sexual dimorphism. In this example, only 30 lengths are recorded.*

Harvest Length											
	Species: Any fish species Date: October 1, 1995 Stocking Date: March 1, 1995 Number stocked: 1,000 fingerlings				Pond: 23 Spawned:April 20,1994 Number offamilies: 27						
Lengths in mm											
	345	354	327	355	330	341	361	357	348	328	
	329	355	359	369	340	351	349	344	348	345	
	352	331	333	336	338	342	347	343	344	355	
mean: to be calculated											

Table: *Example of a data table that can be used to record harvest data for two quantitative phenotypes. This data table is designed for a species that does not exhibit sexual dimorphism for body size. In this example, length and body depth are recorded for only 4 fish.*

Harvest Length and Body Depth Species: Any fish species Date: October 3, 1995 Stocking date: March 4, 1995 No. stocked: 1,230 fingerlings		Pond: 25 Spawned: April 2, 1994 Number of families: 29
Fish No.	**Length (mm)**	**Body depth (mm)**
1	259	127
2	265	135
3	263	133
4	278	139
Means	= to be calculated	= to be calculated

These data tables can be used to record data at any time, from stocking to harvest. In the examples, Tables above were used to record phenotypic values from the sample of fish that were measured to determine the cut-off value(s). The phenotypic values in Tables above would be transferred to the data table illustrated to determine the phenotypic value(s) that corresponds to the desired cut-off percentage.

Table below is a table that takes the means from Tables above at harvest (or at any other time if selection is done prior to harvest) and records the progress that is being made by the selective breeding programme. Mean lengths for the select population and control population, as well as the genetic gain can be recorded in Table.

Tables are examples of data tables that can be used to record data from selective breeding programmes that use family selection. Some of the data tables can be used for both individual and family selection, but those illustrated in Tables are specific for family selection.

Table: *Example of a data table that can be used to record the mean lengths for each generation, and the progress that is made as a result of selection. This data table can be used for both individual and family selection. In this example, only the first two years data are recorded.*

Selective breeding programme to improve growth rate in: Any fish species					
Date	Generation	Original mean	Select population mean	Control population mean	Genetic gain
1994	P_1	315 mm			
1995	F_1		330 mm	321 mm	9 mm

Table: *Example of a data table that can be used to record spawning data for a selective breeding programme that will improve growth rate by within-family selection. The table also includes information on the pond where each family was stocked, the number of fry stocked in the pond, and when the pond was stocked. Ponds 13 and 15 are smaller than the others, so they were stocked with fewer fish.*

Dates: May 1–May 15, 1995 Pond Nos.: 1,2, and 3 Stocked on:April 30,1995					Species: Any fish species Generation: P_1 generation.Fish grown will be selected to become F_1 select brood fish.					
Date	Spawn	Pond	Weight egg mass(g)	Number of eggs	Number hatched	Number of fry	Family	Number stocked	Pond stocked	Dated stocked
5/1	1	1	285	8,550	7,609	6,772	A	200	10	5/7
5/2	2	2	234	7,020	6,458	5,812	B	200	11	5/8
5/2	3	2	27	810	310	175	not used	-	-	-
5/2	4	2	224	6,720	5,644	4,458	C	200	12	5/8
5/2	5	2	174	5,220	3,915	3,719	D	190	13	5/8
5/3	6	3	157	4,710	4,050	3,523	E	200	14	5/9
5/3	7	3	270	8,100	6.053	5,447	F	185	15	6/9

Tables are data tables that can be used to record spawning information for each generation of family selection. Table 20 is for within-family selection, is for between-family selection. The information recorded on these data tables are similar; however, they are organised differently.

The data tables presented in Tables and Figure can be used to record phenotypic values for selective breeding programmes that use family selection. The only difference would be that the data tables would record the family that is being measured.

The data table presented in Table above, can also be used to record yearly means from a breeding programme that uses family selection. If desired, a separate table can be used to record data about every family. Finally, the data tables illustrated in Tables can be used to record the number of fish that spawned each generation.

Table: *Example of a data table that can be used to record spawning data for a selective breeding programme that will improve growth rate by between-family selection. The table also includes information on the ponds where each family was stocked, the number of fry stocked in the pond, and when the pond was stocked. The table is only partially completed.*

Dates: May 5 – May 15, 1995 Pond Nos.: 4,5, and 6 Stocked on: May 1,1995					Species: Any fish species Generation: P_1 generation. Fish grown will be selected to become F_1 select brood fish.					
Date	Spawn	Pond	Weight egg mass (g)	Number eggs	Number hatched	Number of fry	Family	Number stocked	Pond stocked	Date stocked
5/5	1	4	224	6,720	5,690	5,438	A	200	17	5/12
							A	200	23	5/12
							A	200	8	5/12
5/5	2	4	174	5,520	4,055	3,867	B	200	12	5/12
							B	200	20	5/12
							B	200	11	5/12

Table: *Example of a data table that can be used to record the number of fish that are spawned each generation for a selective breeding programme that uses individual and within-family selection. In this example, only two years' data are recorded.*

Selective breeding programme to improve growth rate in: Any fish species						
Date	Number of fish spawned			Number of families used	Number of brood fish Contributing offspring	
	Females	Males	Total		Females	Males
1994	27	27	54	25	25	25
1995	35	25	60	34	34	25

Table: *Example of a data table that can be used to record the number of fish that are spawned each generation, the number of families that are used, and the number of families that are saved for a selective breeding programme that uses between-family selection. In this example, only two years' data are recorded.*

Selective breeding programme to improve growth rate in: Any fish species							
Date	Number of fish spawned			Number of families used	Number of families saved	Number of brood fish contributing offspring	
	Females	Males	Total			Females	Males
1994	27	27	54	26	10	5	5
1995	35	25	60	32	10	5	4

The selective breeding programmes outlined herein earlier demonstrate that growth rate and other quantitative phenotypes can be improved by relatively simple and inexpensive selective breeding programmes. If possible, individual selection should be used because it is the easiest and least expensive. If fish can be spawned synchronously, a selective breeding programme to improve growth rate can be conducted in only one or two ponds, and it will have little effect on normal farming operations. Even if the species cannot be spawned synchronously, if several age cohorts can be produced, the selective breeding programme will still be relatively simple and inexpensive, and it will have minimal impact on normal farming operations. If a farmer cannot control spawning behaviour and if he cannot develop age cohorts or if the phenotype that the farmer wants to improve has a small heritability and is strongly influenced by environmental variables, he must use family selection to improve the phenotype. If possible, within-family selection should be used because it is easier and less expensive than between-family selection. However, the most appropriate type of selective breeding programme is determined by the biology of the species, by the phenotype's heritability, and by the type of environmental factors that influence phenotypic expression, not by a farmer's desire to have an inexpensive breeding programme.

The complexity of a selective breeding programme depends on a number of factors: First and foremost is the number of phenotypes that will be improved. Virtually all selective breeding programmes should attempt to improve growth rate. This is the most important phenotype, because faster-growing fish take less time to reach market and faster-growing fish increase yields. Other phenotypes can be added, but a farmer must add only those traits that are truly important, because the rate of improvement for growth rate will be inversely related to the number of phenotypes that he incorporates into the selective breeding programme. At most, a farmer should add only one or two additional phenotypes.

Secondly, the complexity and cost of the selective breeding programme is determined by the way the fish are produced: a single-phase growing system, where fry are stocked in a pond and grown to market; a two-phase system, where fry are grown to fingerlings in one pond and where fingerlings are then grown to market size in a second pond. If fish are produced under a single-phase production system, selective breeding programmes will be easier and less expensive because fewer ponds will be needed. On the other hand, if fish are produced under a two-phase system, selection can occur twice, which means greater gains can be achieved. Finally, the complexity of the selective breeding programme is partially determined by whether the species exhibits sexual dimorphism for body size. If it does, cut-off values will have to be created for each sex, which means each fish must be sexed in addition to being measured. This doubles the cost of measurement and increases record keeping slightly. One of the end goals of a selective breeding programme should be to save 100–200 select brood fish each generation. This will ensure that a farmer will be able to spawn at least 25 males and 25 females each generation. If this is done, inbreeding-related problems should be minimised for 5 generations. A farmer also needs to save enough select brood fish in order to produce sufficient offspring for the next generation of selection. Although this can be a major concern on large farms, it should be of little concern on medium-sized farms.

If the programme is conducted properly, it can be integrated with and can complement normal farming operations. This is important because if there is a conflict, the selective breeding programme will usually be neglected or abandoned; a farmer's top priority is going to be food production. If there are enough select brood fish, they can be used to produce offspring for both the selection programme and for the production ponds. This will enable a farmer to transfer the genetic gain from the selective breeding programme to the production ponds immediately. If the select brood fish cannot produce enough offspring for both purposes, they must be used to create the next generation of select fish; any surplus offspring can be stocked in the production ponds. In this case, culls from the selective breeding programme or the previous generation's select brood fish (beginning with the F_2 generation of selection) can be used to produce the production fish. This approach will transfer genetic gain from the selective breeding programme to the production ponds, but the transfer will be delayed; in the extreme, the transfer will be with a one-generation delay. This means a farmer must maintain two sets of brood fish: those used to produce the select generation, and those used to produce the fish stocked in the production ponds. Even if the transfer of genetic gain is delayed by one generation, the mean growth rate and yields that are obtained in the selective breeding programme will allow a farmer to predict the growth rates and yields that he will be able to achieve in his production ponds in the future. Additionally, these data will demonstrate that selective breeding is improving his fish and will also enable him to realise that his selective breeding programme is going to create better fish, larger harvests, and greater profits.

Bibliography

Atre, P. K.: *Fish Genetics and Aquatic Environment*, Navyug, Delhi, 2008.

Breder, C. M., and Rosen, D. E.: *Modes of Reproduction in Fishes, How Fishes Breed*, Natural History Press, New York, 1966.

Burgess, Peter: *Tropical Fishlopaedia: A Complete Guide to Tropical Fish Care*, Howell Books, NY, 2000.

Carlton, J.T. and R.I. Smith: *Lights' Manual: Intertidal Invertebrates of the Central California Coast,* University of California Press, Berkeley. 1975.

Chhapgar, B. F.: *Fishes of India*, Oxford University Press, Delhi, 2008.

Deka, Manab: *Fish Fermentation: Traditional to Modern Approaches*, New India Pub, Delhi, 2009.

Dubey, Pushpalata: *Fish Management and Aquatic Environment*, Daya, Delhi, 2006.

Exell A, Burgess PH and Bailey MT: *A-Z of Tropical Fish Diseases and Health Problems*, Howell Book House, New York, 2001.

Ferguson, H.W.: *Systemic Pathology of Fish*, Iowa State Press, Ames, Iowa, 1989.

Grizzle J.M.: *Anatomy and Histology of the Channel Catfish*, Auburn Printing Co, 1976.

Hasler, A. D.: *Orientation and Fish Migration, in Hoar, W. S., and Randall, D. J.*, New York, Academic Press, 1971.

Heda, Nilesh: *Fresh Water Fishes of Central India: A Field Guide,* Vigyan Prasar, 2009.

Helfman G., Collette B., & Facey D.: *The Diversity of Fishes*, Blackwell Publishing, UK, 2004.

Jee, Chandrawati and Shagufta: *Fish Biotechnology*, A.P.H. Pub, Delhi, 2010.

Kulkarni, G.K.: *Fisheries and Fish Toxicology*, APH, Delhi, 2006.

Lewbart G.A. *Self-Assesment Color Review of Ornamental Fish*, Iowa State Press,1998.

Malvee, Sangeeta: *Fish Genetics*, SBS Pub, Delhi, 2008.

Marshall, N. B.: *The Life of Fishes*, Weidenfield and Nicholson, London, 1965.

Moyle, PB and Cech, JJ: *Fishes, An Introduction to Ichthyology*, Benjamin Cummings, NY, 2003.

Nash, Colin (2011) *The History of Aquaculture* John Wiley and Sons. ISBN 9780813821634.

Nelson, J. S.: *Fishes of the World*, John Wiley & Sons, Inc.. , NY, 2006.

Petr, Tomi: *Fisheries in Irrigation Systems of Arid Asia*, Daya, Delhi, 2007.

Pollard, D. A. *Freshwater Fishes of South-eastern Australia*. McDowell, R. M. Reed, Sydney. 1980.

Poppe T.T., *A Color Atlas of Salmonid Diseases*, Academic Press, London, 1996.

Puri, Neelima: *Fish Endocrinology*, Vista International Pub, Delhi, 2008.

Rebelin W.E.: *The Pathology of Fishes*, The University of Wisconsin Press, UK, 1975.

Reichenbach-Klinke H. H.: *Fish Pathology*, T.F.H. Publications, Inc. Neptune City, NJ. 1973.

Roberts, R.J: *Fish Pathology*, Bailliere Tindall, London, 1989.

Robertson, D.R.: *Fishes of the Tropical Eastern Pacific,* University of Hawaii Press, Honolulu. 1994.

Roger C. Helm: *Marine Mammals of California,* University of California Press. 1989

Sandford, Gina: *Aquarium Owner's Guide*, DK Publishing, London, 2000.

Schmida, Gunther E.: *Rainbow Fish*. Barron's Educational Series, NY, 2000.

Sharma, O.P.: *Fisheries Extension and Administration*, Agrotech Pub, Delhi, 2011.

Silva, De: *Fish Nutrition in Aquaculture*, Springer, Delhi, 2001.

Srivastava, C. B. L.: *Fish Biology*, Narendra, Delhi, 2008.

Srivastava, M.P.: *Natural History of Fishes and Systematics of Freshwater Fishes of India,* Narendra Publishing House, Delhi, 2002.

Stoskopf, M.K.: *Fish Medicine*, W.B. Saunders Co. 1993.

Trevor, A. Anderson: *Fish Nutrition in Aquaculture*, Springer, Delhi, 2009.

Yohei, Sakamoto and Mori, Fumitashi: *Aquarium Fish of the World: The Comprehensive Guide to 650 Species*, Chronicle Books, UK, 1991.

Index

❑❑❑